DATE DUE 19724

A Pragmatic Theory of Fallacy

STUDIES IN RHETORIC AND COMMUNICATION
General Editors:
E. Culpepper Clark
Raymie E. McKerrow
David Zarefsky

Douglas Walton

A Pragmatic Theory of Fallacy

The University of Alabama Press   Tuscaloosa and London

∞

The paper on which this book is printed meets the minimum requirements of
American National Standard for Information Science-Permanence of Paper for
Printed Library Materials, ANSI Z39.48–1984.

Library of Congress Cataloging–in–Publication Data

Walton, Douglas
    A pragmatic theory of fallacy / Douglas Walton.
       p. cm.—(Studies in rhetoric and communication)
    Includes bibliographical references and index.
    ISBN 0–8173–0798–2 (alk. paper)
    1. Fallacies (Logic) I. Title. II. Series.
 BC175.W33 1995
    165—dc20   94–23534
                                          CIP

British Library Cataloguing-in-Publication Data available

For Karen, With Love

Contents

Preface

What is a fallacy? Fallacies have been with us since Aristotle, but logic textbooks exhibit little clarity or consistency in giving any helpful answers to this question. This book takes a clear, analytical look at the concept of a fallacy and presents a new, up-to-date analysis of it useful for state-of-the-art argumentation studies. Although many individual fallacies have now been studied and analyzed in the growing literature on argumentation, the concept of fallacy itself has heretofore lacked enough of a clear meaning to make it as useful as it could be as a tool for evaluating arguments.

The view put forward is one that will appear radical and controversial to traditionalists in logic. A fallacy is regarded as an argumentation technique, based on an argumentation scheme, misused to block the goals of a dialogue in which two parties are reasoning together. The view is a pragmatic one, based on the assumption that when people argue, they do so in a context of dialogue, a conventionalized normative framework that is goal-directed. This framework is crucial in determining whether the argumentation used is correct or incorrect, in relation to the given details of a specific case, according to the account given here.

Of course, this pragmatic and dialectical type of view of fallacy is already familiar from the writings of the Amsterdam School, who define a fallacy as a violation of a rule of a critical discussion. This account of fallacy has the advantage of being a pragmatic definition that views fallaciousness in relation to the context of dialogue in which an argument was used. In this book, however, it is shown why this view defines only a necessary condition, or an approximation thereto, and not a sufficient definition of the concept of fallacy. More needs to be added to it, to make it the basis of a useful program for the analysis of fallacies. It is argued that the Amsterdam definition of fallacy falls short of drawing the important distinction between a fallacy and a blunder in argumentation, the latter being a less serious type of error.

The Amsterdam view is limited to just one type of dialogue that is vitally important in defining the concept of fallacy—the critical discussion. In this new view, the analysis is extended to several other key types of dialogue that can also serve as normative models of argumentation. But six basic types are shown to be of key importance—the critical discussion, the inquiry, negotiation, deliberation, the quarrel, and information-seeking dialogue. An important feature of this multiple approach is that it shows how fallacies can often be analyzed as illicit dialectical shifts in argumentation from one type of dialogue to another.

The analysis of the concept of fallacy presented shows that there is no one-to-one defining correspondence between a fallacy of a particular type and a violation of one of the rules for a critical discussion given by van Eemeren and Grootendorst. Unfortunately for simplicity, the job of defining the concept of fallacy turns out to be more difficult than this proposal allows.

Three problems confronted are that of fallacy identification, fallacy analysis, and fallacy evaluation. All three problems are solved by developing new pragmatic structures that display the form of an argument (the so-called argumentation scheme) and then show how this form fits into an enveloping normative structure of dialogue. In this book it is shown how the twenty-five or so major informal fallacies of the standard treatment in the textbooks are basically reasonable presumptive types of arguments that have been used inappropriately in such a normative model.

Chapter 5 presents twenty-five of these basic argumentation schemes for presumptive reasoning, with sets of accompanying critical questions matching each scheme. The view of fallacy then presented is that a fallacy is either a *paralogism*, an argumentation scheme used in such a way that it systematically fails to answer a critical question appropriate for that scheme, or a *sophism*, a more extended misuse of a scheme, or sequence of them connected together, that has been twisted or used incorrectly in a dialogue, as evidenced by a distorted profile of dialogue or what is called an argumentation theme. A theme is a sequence of connected moves in a dialogue, displayed by a profile, or tableau, a pair of matching columns of displayed moves. A profile that shows evidence of a fallacy having been committed is one that has been distorted, or "balled up." The moves are in an order or pattern that is normatively inappropriate in just this sense. The moves occur in a structurally blocking or interfering order with respect to forwarding the goal of the dialogue.

The paralogism type of fallacy is a systematic, underlying type of error of reasoning in an argument. The sophism type of fallacy occurs where a scheme is used as a deceptive tactic to try to get the best of the other party unfairly when two parties reason together in one or more of the several types of dialogue.

The second chapter summarizes the standard accounts of the twenty-five or so major fallacies treated. The fourth chapter identifies the goals and techniques of argumentation characteristic of six basic types of dialogue, presented in the book, in which argumentation is said to occur. Other chapters are on relevance, on formal fallacies, and on how the twenty-five or so major, traditional fallacies support the new theory.

The book shows how the examples of the fallacies given in the logic textbooks characteristically turn out to be variants of arguments that are reasonable, even if defeasible or questionable, and are based on presumptive reasoning. This is the essence of the evaluation problem. It is a key thesis of the book that we must not take for granted, as the textbooks in the past have so often done, that a fallacy may be spotted simply by looking at the type of argument it is, apart from how it was used in a context of dialogue. This point is shown to be especially important when dealing with the sophistical tactics type of fallacies, where the profile of dialogue is all-important in showing how the argument was used, for example, too aggressively, to bring undue pressure to bear on a participant.

It is argued in the book that questionable arguments, and blunders in argumentation, need to be distinguished from fallacious arguments. It is stressed that the claim that an argument is fallacious should be seen as a strong form of condemnation that needs to be backed up by certain kinds of evidence that meet a burden of proof appropriate for such an allegation.

Formal logic has been successful as a scientific discipline because it is based on argument forms that can be evaluated as valid or invalid in an enveloping structure. What has been lacking, however, is an informal or practical logic that judges the use of an argument in a given case, in a context of conversation. The intent of this book is to contribute to a restoration of this imbalance by basing a theory of fallacy on argumentation schemes that can be evaluated on the basis of their use in different developing structures of dialogue.

I developed the ideas on which the book is based and wrote the first drafts of some chapters during two periods when I was a fellow-in-residence of the Netherlands Institute for Advanced Study in the Humanities and Social Sciences. My thanks are due to the staff and fellows of NIAS in 1986–87, and again in 1989–90, for their kindness, for helping a foreign scholar to feel at home in Holland, and for providing a stimulating and charming climate that eased the toils of writing and thinking during this period of intense work. I am especially grateful to Erik Krabbe, whose daily discussions on argumentation at NIAS were a continuing source of stimulation and insight that provoked new ideas and refined old ones.

The research that culminated in this book was supported by three sources: (1) a Killam Research Fellowship from the Killam Program of the Canada Council, (2) a fellowship from the Netherlands Institute for Advanced Study in the Humanities and Social Sciences, and (3) a research grant from the Social Sciences and Humanities Research Council of Canada. I am also grateful to Dean Michael MacIntyre and the Department of Philosophy of the University of Winnipeg for making arrangements that enabled me to be absent from my department during the period of my stay in NIAS.

I thank Paul van der Laan for supplying etymological information on the origins of the word 'fallacy,' Erik Krabbe for pointing out the distinction between a *Fehlschluss* and a *Trügschluss*, and Siegwart Lindenberg for clarifying the usage of these terms.

For discussions on related matters during 1989–90 at NIAS, I thank Frans van Eemeren, Rob Grootendorst, Scott Jacobs, Sally Jackson, Agnes Verbiest, Agnes Haft-van Rees, Charles Willard, and John Woods. Finally, for reading through the manuscript at the later stages, and making many corrections, I thank Hans Hansen and Erik Krabbe.

Special thanks are due to Amy Merrett for the information processing of the text and figures of the manuscript through the many drafts. Finally, I would like to thank Rita Campbell for making the index and Harry Simpson for help with proofreading.

It is as the enemy of Fallacy that Logic must always find its application to real life: Fallacy occupies much the same position in regard to science of Proof that disease occupies in regard to the science of Medicine.

—Alfred Sidgwick, *Fallacies: A View of Logic from the Practical Side*

1

The Concept of Fallacy

In the growing literature in the field of argumentation, the single area that has been most intensively researched in recent years is that of fallacies. Fallacies are portrayed in the new view put forward in this book as important, baptizable[1] types of errors or deceptive tactics of argumentation that tend to fool or trip up participants in argumentation in various kinds of everyday discussions. One problem is that many of the individual fallacies have been studied and carefully analyzed, but there remains concern as to whether the concept of fallacy itself is itself clear or coherent enough to sustain its central place in the field of argumentation. The general presumption in the field of informal logic is that although the concept of fallacy is here to stay, and is too important to dispense with or ignore, it lacks enough of a clear, underlying structural basis to make it useful as an analytical tool to help with the systematic evaluation of arguments.[2]

1. Greek Roots of the Concept of Fallacy

The original Greek idea of a fallacy, found in Aristotle's practical manual on the art of argumentation, the *De sophisticis elenchis* (*On Sophistical Refutations*), viewed a fallacy (or sophistical refutation) as a deliberate deceptive tactic of argumentation used to trick and get the best of a speech partner in dialogue unfairly. But this idea afterward fell into disuse and along with it the background framework of practical logic as a dialectical art of conversation between two parties who reason together. In its place, Aristotle's syllogistic logic, and with it the idea of deductive logic as a system for testing inferences for validity, took over as the dominant point of view in logic. The view of fallacy that evolved into the modern logic textbooks took on this dominant point of view, seeing a fallacy as an erroneous inference—a kind of error of reasoning that was a faulty inference from a premise to a conclusion.[3] This viewpoint abstracted away the concept of argument as an exchange in dialogue between two parties.

Recent research on the fallacies is now revealing the limitations of this modern viewpoint.[4] While some fallacies can be quite usefully analyzed as errors of reasoning or faulty inferences, we clearly cannot make sense of many of the major informal fallacies unless we revert to something like the Aristotelian conception of logic as a dialectical art.

As Hamblin showed in detail, the basic problem with fallacies is that the Aristotelian classification of them has been handed down in the logic textbooks basically intact for over two thousand years, but the classification makes little sense to modern readers because we have lost the underlying idea of fallacy.

The incoherence of the concept of fallacy for two thousand years was compounded by linguistic problems. The Greek concept of *elenchus* (refutation), meaning reasoning involving the contradictory of a given conclusion, seems mysterious and alien to modern readers (Hamblin 1970, 50). As Hamblin also notes (50), Greek has no precise synonym for 'fallacy,' and the two main terms used by Aristotle, *sophistikos elenckos* (sophistical refutation) and *paralogismos*, are often translated as *sophism* and *paralogism*. This practice often makes for confusion, because the term 'sophism,' for example, is often used to refer to inconsistencies, paradoxes, or other forms of logical anomalies that are quite different from the kinds of phenomena that tradition identifies as informal fallacies.

In certain ways, Aristotle's philosophy of sophistical refutations turned out to be alien and incomprehensible to subsequent generations of readers (especially since the advent of mathematical logic). For it was set in a framework of different types of argument in discussion, each with its own distinct goals, based on the presumptions that arguments can start from generally accepted opinions and that their mode of operation is a sequence of exchanges between a questioner and a respondent where each party has goals of argument, like refutation of the other party. See Kapp (1942) and Evans (1977). Thus the concept of fallacy as the use of an instrumental technique for carrying out goals in an argumentative interpersonal exchange—one that has fallen short of the right way of realizing a goal—was familiar to the Greek conception of applied logic, but it was not a concept that survived in the treatments of the modern logic textbooks. In chapter 8, a new analysis to replace this ancient concept of a fallacy is given—an analysis that is stated in the context and language of recent developments in argumentation theory.

Aristotle introduced the list of methods of arguing that have subsequently been identified with the famous informal fallacies of the logic textbooks. These techniques were called "modes of refutation" by Aristotle (1955, 165 b 23). It seems that this list of techniques were regarded as means or instruments to carry out the goals of argument listed above but especially the goal of refutation. Although Aristotle discussed them in his chosen context of "competitive and contentious" arguments, it appears to be quite possible that these techniques could also be used as methods in other contexts of argument as well. That is, although Aristotle was concentrating on the sophistic arts and fallacies, it appears quite possible that these same techniques of argument could also be used to carry out quite legitimately the aim of refutation in one of the other contexts of argument or in contentious argument.

A point noted by Hamblin (1970, 51) is that Aristotle was clearly writing about deliberate sophistry when he discussed fallacies. And in the *De sophisticis elenchis* (174 a 16), Aristotle devoted considerable attention to explaining how these so-called fallacies or sophistical refutations can be used very effectively as techniques or tactics to trick or defeat an opponent in argumentation. This practice is all the more disorienting to the modern reader schooled in the tradition of looking for "fallacies" that are invalid inferences, abstracted from any context of dialogue. The basic problem for informal logic as a scientific discipline is that forms of argument corresponding to the types of arguments used to commit the fallacies have never been identified as well-defined structural units, in the way that forms of argument have been identified and defined in formal logic.

In chapter 5, this key obstacle is overcome by identifying twenty-five or so argumentation schemes representing common types of argumentation used in presumptive reasoning in a context of dialogue to support conclusions. The schemes by themselves are not sufficient to analyze and evaluate the fallacies, however. It has been apparent for some time that the schemes need to be evaluated in a contextual setting of dialogue structures, where they are used (Walton 1989a).

2. Informal Logic as Dialectical

Aristotle's syllogistic, along with propositional logic of the kind studied by the Stoics, evolved into the modern formal logic, used to determine deductive validity of arguments. But logicians have long regarded the idea of judging nonconclusive arguments, as used in a conversational context, with a lack of interest or even with suspicion. The undeveloped state of the "fallacies" sections of the logic textbooks attests to an unwillingness by logicians to attempt to evaluate argumentation seriously in this informal, practical, or applied manner. Why is this so? The problem seems to be that informal logic is identified with strategies of persuasion where two parties reason together. To Western logicians, this identification has seemed to come uncomfortably close to rhetoric and salesmanship.

The danger of integrating logic with rhetoric by studying the argumentation tactics involved where premises are opinions and conclusions are drawn by presumptive inferences has been readily apparent in Western philosophy since Plato. Although Plato's philosophical method was that of the Socratic dialogue, he strongly denounced the Sophists precisely because they claimed expertise in knowing how logos is able to operate on opinion with an enormous power of persuasion. Plato denounced "mere opinion" as inherently misleading as a premise base for argument that yields real insight, and those who based their reasoning on it as unreliable and dishonest purveyors of fallacies who have no respect for the truth and use their rhetorical methods strictly for personal profit. Ever since this Platonic denunciation, exponents of opinion-based reasoning have been consistently rejected in Western thought, and 'rhetoric' has come to stand for "cheap talk," colorful and emotional speech-making that is the diametrical opposite of logical reasoning, scientific method, or plain talk motivated by an honest concern for the truth of a matter.

This belittling of opinion-based reasoning as something that is very much of secondary importance in serious intellectual undertakings (like science and philosophy) peaked in Descartes's method of doubt, which required adopting only premises based on certain and indubitable knowledge. The subsequent successes of the empirical and mathematical sciences after Descartes naturally led to the development of a mathematical (symbolic) logic of propositions based on truth values and truth functions. By the beginning of the twentieth century, Plato's denunciation of the possibility of any *techne logon*, or skill of logic (informal logic), based on opinion and presumptive inference had become solid orthodoxy, excluding any serious investigation of this area as a respectable part of the science of logic.

The dominance of the semantic, deductivist conception of reasoning in logic is often attributed to the rise of mathematical (symbolic) logic in the twentieth century. But the roots of this dominance, and the underlying climate of intellectual opinion that gave rise to it, go much deeper. Analytical philosophy in the twentieth century has tended to concentrate on experimental and mathematical science as the paradigm of logical reasoning. This attitude stems from the rise of the scientific method since the Renaissance, which has also seen a progressive decline in the humanities. Traditional dialectical methods of the humanities, which stressed the interpretation of texts of discourse, and informal methods to minimize bias and promote empathy (the ability to see both sides of an issue), came to be perceived as skills with little or no market value and as "artistic" or "literary" undertakings to be firmly excluded from science.

But the roots of these attitudes go even deeper. The semantic conception deals with "hard," "objective" matters of truth values and deductive validity. Any foray outside these narrow borders tends to be greeted with contempt because it appears to take us into the realms of acceptance and subjective opinion. And as any student relativist will tell you, in these regions, my opinion is just as good as yours. The prevailing attitude is that there can be no objectively justifiable reason for claiming that one person's opinion is better (more right) than another's. Hence the popularity recently of the deconstructivist type of approach that disparages the possibility of any kind of rationality, as applied to the important affairs of everyday life.

According to this point of view, arguments based on burden of proof in a balance of considerations and argumentation schemes could never have solid verifiable validity. By making the evaluation of arguments as correct or incorrect a matter of the use of presumptive argumentation schemes used in an interpersonal context of dialogue, many critics will feel, logic has been cast not only into a kind of relativism to a context of dialogue (pluralism), but also into a sophistic equation of the persuasive argument that works (the good tactic) with the correct or logically sound argument. No doubt, many critics will feel that such a theory is a disaster for logic, a corruption of logic as an objective science of argument evaluation.

But this criticism presupposes the widely accepted point of view that logic is a science of *monolectical* (not dialectical) reasoning that evaluates a set of propositions as valid or invalid independently of the context of dialogue in which this reasoning was used or put forward and that logic is a science of *monotonic* reasoning that is concerned with fixed and unchanging truth values of these propositions, unaffected by the changing status of presumptions in a dialectically fluid situation. This monolectical-monotonic framework is not very useful, however, for addressing the informal fallacies.

The main problem is that in the twentieth century we have become accustomed to think of logical reasoning in a monotonic and monological framework of relationships exclusively concerned with given relations on values of truth and falsity of a set of propositions. Aristotle's treatment of fallacies, as noted already, however, presupposed a dialectical framework of two persons making a sequence of moves in presenting arguments to each other in an organized exchange of viewpoints. Hence it is not hard to see why this subject has languished in such a neglected state for so long.

The traditional monological-monotonic framework that was orthodoxy for so long tended to see "fallacy" only as a failure of validity, adding a psychological concept of "seeming to be valid" as an afterthought. Aristotle's whole treatment of the fallacies, however, was deeply dialectical. And the attempt to view his list of fallacies from a monolectical-monotonic point of view made them appear to be either trivial or incomprehensible as objects of study in logic.

Hamblin (1970, 66) brought out this point very well when he wrote, "In our attempt to understand Aristotle's account of fallacies we need to give up our tendency to see them as purely logical and see them instead as moves in the presentation of a contentious argument by one person to another." This reorientation to a dialectical way of conceiving argument means coming to think of the fallacies as means or instruments used by one participant to carry out objectives in a two-party sequence of exchanges where each party has a goal in the dialogue. Of course, it is especially important to understand these instrumental strategies of interpersonal argument because they can be used to trick and deceive the other party in contentious debates. But because they are instruments, the possibility is there that they can be used to achieve legitimate ends of dialogue without deceit or error, even if the conclusions drawn are provisional and relative to the purpose of a conversation.

Traditionally, however, and even continuing into the current logic textbooks, dialectical arguments are seen purely as fallacies, with little or no acknowledgment of their positive or correct side. This negative approach emphasizes the deceitful and erroneous aspect of dialectical argumentation.

3. State of the Art of Dialogue Logic

Looking at the modern logic textbooks, we see that the names of many of the traditional fallacies are derived from the same terms used by Aristotle to describe sophistical refutations.[5] What is lacking is any kind of framework in which to place the concept of fallacy that is anything like the Aristotelian conception of an interpersonal exchange of reasoning in which sophistical refutations are used as techniques of argumentation to trick or deceive a partner in dialogue.

Hence we have arrived at a point in history where analyses of the individual fallacies cannot go much further without our taking a hard look at the conversational framework in which arguments are used. Fortunately, we are not entirely without resources for undertaking such a project, for in recent times, there has been a return within the informal logic movement toward looking at an argument as a dialogue exchange between parties who reason together. Indeed, there have even been proposed formal frameworks of structures of dialogue as contexts for argument. And in addition, there have been studies on the pragmatics of argumentation in the field of speech communication. These developments, outlined below, all point to a revival of the Greek idea of practical logic as a dialectical art of reasoned conversation, where arguments are exchanged between two parties.

At the present state of the art, there are two different kinds of approaches to formulating sets of rules of reasonable dialogue. One is the formalistic approach of devising sets of rules for abstract games of dialogue designed to model or approximate argumentative discussions (Hamblin 1970; 1971). Hamblin's methods have been pursued by Mackenzie (1981; 1990) and Walton (1985a; 1987). Hintikka (1981) first constructed games of this sort to model questioning but then later (1987) applied them to the topic of fallacies as well. Independently, Lorenzen (1969) constructed formal games of dialogue, and these games have been applied to argumentation and fallacy by Barth and Krabbe (1982).

The other approach comes from recent research in the field of speech communication. It is less formalistic and more practical in nature. While this approach is certainly compatible with formalization, it could be more generally categorized as pragma-dialectical. This type of approach formulates general rules that support a goal of a particular type of dialogue—these rules are linguistic (pragmatic) rules for speech acts, and they are stated in natural language. The general approach to rules is based on the conversational maxims of Grice (1975), implicit rules that function as conventions of politeness upheld by participants in a cooperative conversation. The kind of conversation van Eemeren and Grootendorst (1984; 1992) describe as the normative framework for reasoned argumentation is called the critical discussion. A critical discussion is a type of dialogue in which there are two participants who have a conflict of opinion. The goal of the critical discussion is the resolution of this conflict of opinion through argumentation, according to the rules appropriate for the critical discussion. The critical discussion is a normative model of reasonable (good, ideal) dialogue against which texts of argumentative discourse can be evaluated. A fallacy, for van Eemeren and Grootendorst (1984), is an incorrect move, a move that violates the rules of a critical discussion: "These incorrect moves correspond roughly to the various kinds of defects traditionally referred to as *fallacies."* (284). This approach is quite different from the traditional one, which sees a fallacy as an invalid inference. This approach sees fallacies as failures of communication—failures to conform to conventions necessary to carry on a conversation.

The traditional approach had been to attempt to formalize arguments individually and then to show how the apparently valid ones are not actually so. As van Eemeren and Grootendorst (1989) pointed out, much of the early work on fallacies done by Woods and Walton, published in the collection of papers (1989), took this approach of applying different nonstandard logical systems to the fallacies to pinpoint the error as an incorrect or invalid type of inference. Van Eemeren and Grootendorst (1989, 102) saw the work of Walton (1987) as a turning point or transitional stage to a pragmatic and dialectical approach that sees fallacies as errors, deceptive tricks, or failures in how argumentation is used in a context of dialogue where two parties reason together. Van Eemeren and Grootendorst heralded this new approach, noting that Hamblin himself (1970) had advocated that games of dialogue, where two (or several) parties exchange information and arguments, provide the right setting for analyzing the fallacies.

Hamblin himself had constructed several games or formalized models of dialogue.[6] And even in the most elaborately formalized of these—the 'Why-Because-Game-with-Questions'—the ultimate goal or purpose of the game was left vague and open-ended (no doubt purposely). The dialogue was said to be "information-oriented" (Hamblin 1970, 271). But exactly what counts as "information" was left open. This openness has been worrisome to many interested in using games of dialogue to analyze fallacies. Does it allow a pluralism of all kinds of games of dialogue? Can you simply invent formal games of dialogue at will and declare them legitimate? The problem is that there do seem to be different uses of argumentation in different types of dialogue that have different goals. What is needed is an analysis of several of these types of dialogue—at least the ones that are most prominent in studying the contexts of argumentation of the major informal fallacies. Although formal dialogues in the Hamblin style have been constructed by Mackenzie (1981; 1990), the pluralism of these structures leaves open the question of how they might be used to define or clarify the concept of fallacy.

Hintikka (1987) also proposed a shift in our approach to fallacy that leads in the same direction. He argued that it is best to reject the traditional assumption that fallacies are invalid arguments and replace it with the idea that fallacies are breaches of proper procedures of question-answer dialogue. This proposal is both a turn toward the pragmatic and also a return to the kind of framework of question-reply dialogue that Aristotle presupposed in the *Topics* and *De sophisticis elenchis.*

The present book argues that these approaches to contextualization of argument have been, in key respects, either too broad or too narrow. The pragmatic approach of having only one normative model that is the proper context of argument for analyzing fallacies, like the critical discussion, is too narrow. For, as we will see below, fallacies and argumentation can also occur in other contexts of dialogue, like negotiation, for example, that are distinctively different in structure from the critical discussion. The Hamblin-style approach of constructing formalistic models of dialogue reasoning has been too broad, because it has led to a proliferation of formal systems that are precisely enough motivated to model the different purposes of the distinctive types of conversational frameworks in which argumentation takes place.

To meet the right level of analysis needed for the fallacies, several goal-directed normative models of dialogue are analyzed in chapter 4. These types of dialogue provide structures that can help us evaluate how an argument is used correctly or incorrectly when two parties reason together and they are of the right level of generality to be helpful with the project of analyzing fallacies.

According to the new approach set out in this book, a fallacy can be a violation of a rule of a critical discussion, or a violation of a rule of a type of dialogue other than a critical discussion, or in some cases it can even be an illicit shift from one of these types of dialogue to another. In a word, the new theory of dialogue, put forward in chapter 4 later, is *pluralistic*—it postulates several different normative models of reasonable dialogue. Hence the use of the term 'dialogue,' which is meant to be more inclusive than the term 'critical discussion.'

In this new theory of fallacy, however, the normative models of dialogue that are appropriate frameworks for argumentation will not be arbitrary, or purely formal, structures. Six of the most common and typical types of dialogue will be fully defined as normative models in enough detail so as to be practically useful in the project of analyzing and evaluating the fallacies.

The new concept of fallacy postulated in this book is more complex than that of the Amsterdam School because a fallacy is no longer defined as just a violation of a rule of reasonable dialogue. Some violations of rules of dialogue will be classified as flaws, blunders, or errors that are not so bad or serious that they are classified as fallacies. A sophistical tactics fallacy—the type of fallacy on which the analysis in this book will concentrate—will be shown to be a special kind of violation of the rules where a baptized type of argumentation technique has been misused as a tactic or deceptive trick by one participant in order to try to get the best of the other participant in the dialogue. The resulting new theory of fallacy identifies a fallacy with the means, the type of argument, that was used to violate the rule. The new concept of fallacy is formulated in relation to systematic kinds of wrongly used argumentation techniques in several key contexts of dialogue in which argumentation occurs in everyday conversations.

4. Fallacies and Violations of Rules

It is tempting to think that the rules for a critical discussion can be used to classify a particular fallacy as a violation of a particular rule, so that when a specific fallacy occurs, you can say, "There, that was a violation of rule x, therefore it is a case of fallacy y." But the rules for the critical discussion given by van Eemeren and Grootendorst (1984; 1987) are very broad. They express in broad terms guidelines on the means to carry out the goals of the dialogue. For example, there is a rule that burden of proof must be fulfilled when requested. But as we have seen, most, if not all, of the major fallacies involve failure to fulfill this requirement. And it is exactly how such a failure occurs, by what means, that defines which fallacy occurred, or whether a fallacy occurred. For failure to fulfill burden of proof is not itself a fallacy at all, much less any specific fallacy. What needs to be determined is what specific means were used to carry out the failure.

The means are basically arguments, types of argumentation that must be used in certain ways, if the goals are to be achieved. Specific ways of misusing these arguments, ways that block the goals, are fallacies. But there is no one-to-one correspondence between the rules and these fallacies. Unfortunately for theorists, the fallacies turn out to be more complex than that.

The ten rules for critical discussion of van Eemeren and Grootendorst (1987) do not identify individual fallacies, but as indicated in Walton (1989a, chap. 1), they do give broad guidelines that yield definite insight into what is basically wrong when fallacies are committed.

According to van Eemeren and Grootendorst (1987), the ten rules for the conduct of a critical discussion are the following.[7]

Rule 1:   Parties must not prevent each other from advancing or casting doubt on standpoints (284).

Rule 2: Whoever advances a standpoint is obliged to defend it if asked to do so (285).

Rule 3: An attack on a standpoint must relate to the standpoint that has really been advanced by the protagonist (286).

Rule 4: A standpoint may be defended only by advancing argumentation relating to that standpoint (286).

Rule 5: A person can be held to the premises he leaves implicit (287).

Rule 6: A standpoint must be regarded as conclusively defended if the defence takes place by means of arguments belonging to the common starting point (288).

Rule 7: A standpoint must be regarded as conclusively defended if the defence takes place by means of arguments in which a commonly accepted scheme of argumentation is correctly applied (289).

Rule 8: The arguments used in a discursive text must be valid or capable of being validated by the explicitization of one or more unexpressed premises (290).

Rule 9: A failed defence must result in the protagonist withdrawing his standpoint and a successful defence must result in the antagonist withdrawing his doubt about the standpoint (291).

Rule 10: Formulations must be neither puzzlingly vague nor confusingly ambiguous and must be interpreted as accurately as possible (292).

A fallacy is then defined by van Eemeren and Grootendorst (1987, 284) as an incorrect move in a critical discussion—'incorrect' in the sense that it violates one or more of these rules. Van Eemeren and Grootendorst (284) add that the term 'fallacy' in their theory refers to a move in argument that "hinders the resolution of a dispute in a critical discussion."

Rule 1 applies to many of the cases of fallacies, because different tactics are often used to try to prevent parties from expressing or casting doubt on a standpoint. Van Eemeren and Grootendorst themselves concede that *ad hominem, ad baculum,* and *ad misericordiam* (212–13) all violate this rule. So this rule does not single out any particular fallacy.

Rules 3 and 4, in effect, stipulate that an argument must be relevant to the issue of a dialogue. Although irrelevance could be called one big fallacy of *ignoratio elenchi* (thus violating these two rules), many of the fallacies, are in significant part, but not totally, characterizable as failures of relevance.

Rule 2 expresses the idea of burden of proof. But failure to defend your thesis if requested to do so is not, in itself, a fallacy (as noted above). Nor is it identifiable with any single fallacy. Failure to back up one of your contentions when you are asked to do so is a fault or error—it means your argument does not meet the burden-of-proof requirement and is therefore unsupported or insufficiently proven. But that, in itself, does not mean the argument is fallacious. Many of the fallacies are associated, at least in part, by a failure to fulfill burden of proof. Begging the question is one; ad hominem is another (Walton 1985a; 1990a).

The fallacy most intimately connected with failure to fulfill burden of proof is the *argumentum ad ignorantiam*. But the fallacy here is the inappropriate shifting of the burden of proof onto the other party. The fallacy is not itself identical to the failure to fulfill burden of proof. It is not just a violation of rule 2, and that's how it is defined as a fallacy. It is a special type of tactic used to try to shift burden of proof deceptively or inappropriately from one side to the other in a dialogue (Walton 1992d).

The same can be said for the fallacy of *petitio principii*. Although sometimes wrongly identified with the fault of failure to fulfill burden of proof, such a failure is not identical or equivalent to the fallacy of *petitio*. The fallacy of begging the question, or petitio principii, involves the essential use of circular argumentation that, while not fallacious in itself, is fallaciously used to evade a proper fulfillment of burden of proof in a dialogue (Walton 1991a).

Rule 7 (and possibly with it rule 8) is a sort of granddaddy rule that covers most of the major informal fallacies (rule 8 perhaps covering the formal fallacies, depending on what is meant by 'valid'). For as we have seen, most of these fallacies are essentially arguments where there has been no defense by means of an appropriate argumentation scheme correctly applied. This rule, then, like the others, does not equate with any single fallacy. Rather, its violation can be partially identified with many of the fallacies in some cases. It is not a characteristic, or a defining condition, or an analysis, of any single fallacy.

But the rules are connected to the fallacies. The rules give you a broad insight into what went wrong with a particular fallacy with respect to its getting away from supporting the goals of a dialogue. For example, the rule "Be relevant!" can be used to explain why a particular argument that was wildly off topic by making a personal attack in the midst of a scientific inquiry is blocking the dialogue from taking its proper course. Or the rule "Fulfill the burden of proof!" may indicate what's wrong when someone keeps attacking the other party personally with wild innuendo without backing it up by any real evidence of wrongdoing. But in both cases, the fallacy might be an ad hominem fallacy. Here the rule gives you insight into what has gone wrong basically, but it does not pinpoint or identify the fallacy. It might be quite misleading to say that both these cases are instances of the ignoratio elenchi fallacy, when it would be much more specific and accurate to say that they are instances of the ad hominem fallacy.

As shown by Walton (1989a, chap. 1), there can be more general and more specific rules of dialogue, and there can be positive and negative rules. The goals link to general rules, which in turn link to more specific subrules for specific situations, and in turn, these subrules link more closely to fallacies that occur in certain kinds of arguments.

The following is a good example. In a critical discussion, the goal is to resolve a conflict of opinions by giving each side the freedom and incentive to bring out its strongest arguments to support its side of the issue. For the critical discussion really to succeed, there must be a clashing of the strongest arguments on both sides and good responses by the other side when an argument is very telling against its point of view. In turn, for this to happen, there must be freedom on both sides to express one's point of view as fully as possible. Clearly, in order to function successfully, a critical discussion needs freedom to express a point of view. A certain quality of openness on both sides is required. This positive requirement leads to a negative rule, namely the rule that neither side must prevent the other side from expressing its point of view.

There are all kinds of ways of violating this negative rule, however. One party may say "If you know what's good for you, you will shut up right now!" Or one party may ask an unfair question like, "Have you stopped your usual cheating on your income tax?" Or one party may simply keep talking, out of turn, thus preventing the other party from saying anything at all. The first two tactics are two different types of fallacies, and the third is not a fallacy at all—or at least it is not specifically identifiable with any of the traditional list of fallacies. The first tactic is to make a threat that will presumably prevent the other party from putting forward any further argumentation at all. The second tactic poses a question such that, no matter which answer the respondent gives directly, he or she concedes a defect of veracity that prohibits him or her from putting forward any further arguments that will have any credibility in the discussion.

To make the first two fallacies fallacies that are so by virtue of being specific rule violations, we must invent rules like the following: "Don't make threats in a critical discussion, or in any other type of dialogue where making such a threat necessarily prevents the other party from taking part properly in the dialogue!" or "Don't ask complex questions with presuppositions that are defeating to the respondent's side, unless you get her to agree to the presuppositions using prior questions in the appropriate sequence of dialogue first!" These very specific rules are now sharp enough to characterize particular fallacies.

It doesn't seem to be these kinds of specific, yet qualified rules that van Eemeren and Grootendorst have in mind, however. And moreover, it is obvious that no matter how you define or characterize a particular fallacy, once you have the characterization of it, you can always make up a more complicated rule saying, "Don't do that!" But this way of proceeding would be a circular way of saving the definition of fallacy as a violation of some set of rules for argumentation in dialogue. In general, then, pointing to a rule-violation is not a sufficient way of either pinpointing that a particular fallacy was committed or of evaluating that argument as fallacious. We have to look elsewhere both to identify the fallacies and to define each of them as distinctive entities.

5. The New Approach to Fallacies

The new concept of fallacy is by no means altogether "new" in the sense that, at least in broad outline, it represents a revival of, or a return to, the spirit of the "old" Aristotelian concept of the sophistical refutation. The return is more to the spirit than to the letter of Aristotle, however. The new theory turns out to be quite different from Aristotle's approach in many ways, and it is expressed within the framework of state-of-the-art twentieth-century developments in logic and discourse analysis.

Lambert and Ulrich (1980, 24) claimed that the study of informal fallacies is a questionable enterprise because even after one learns to recognize alleged examples of the various "fallacies," it is difficult to see what common factor makes them all instances of the *same* fallacy. This criticism of the traditional treatment of fallacies is quite accurate. It has long been known that there are borderline cases that could be this fallacy or that fallacy. This is the problem of identifying fallacies.

This book makes it possible to solve the identification problem of fallacies by identifying the argumentation schemes that define the type of argumentation corresponding to many of the various individual fallacies. This method by itself, however, does not solve the problem completely, because (1) the relationship between the argumentation schemes and the individual fallacies is more complicated than a simple one-to-one relation and (2) some of the fallacies do not relate to distinctive argumentation schemes in the way that others do. The identification for this second class of fallacies is solved by introducing argumentation themes, characteristic profiles of sequences of dialogue.

The analysis problem is solved using the argumentation schemes, themes, and the types of dialogue as structures to identify the patterns of argumentation characteristic of the various fallacies as uses of argumentation that fail to be correct, but nevertheless seem to be plausible, because of underlying shifts from one context of dialogue to another.

The new approach is pragmatic—a fallacy is an argumentation technique that is *used* wrongly in a context of dialogue. Fallacies are not arguments per se, according to the new theory, but *uses* of arguments. A fallacy doesn't have to be a deliberate error in a particular case, but it is a question of how the argumentation technique was used in that case. The new theory is not a psychologistic theory but a pragmatic theory. It is a rich explication of the concept of fallacy as a calculated tactic of deceptive attack or defense when two people reason together in contestive disputation. So conceived, a fallacy is not only a violation of a rule of a critical discussion but a distinctive kind of technique of argumentation that has been used to block the goals of a dialogue, while deceptively maintaining an air of plausibility, either by using a type of argumentation that could be correct in other cases or even by shifting to a different type of dialogue illicitly and covertly.

According to the new theory, a fallacy is an underlying, systematic error or deceptive tactic. Charging someone with having committed a fallacy in his argument is quite a serious charge in matters of conversational politeness. It is a serious charge, and it calls for a serious reply, if the alleged offender is to maintain credibility as a serious proponent of his side of the issue of a discussion. A fallacy, then, is not just any error, lapse, or blunder in an argument. It is a serious error or tricky tactic, and its exposure destroys the argument if the offensive move is not corrected or retracted.

Moreover, a fallacy is not just a weak argument that has not been strongly enough backed up by sufficient evidence. The term 'fallacy' refers to an underlying systematic error or misdemeanor in the structure of an argument, a basic flaw indicating that the argument is fundamentally flawed in some way. A fallacy, therefore, is not just any error or violation of a rule of critical discussion that occurs in an argument. It is a serious kind of underlying failure in the way the argument was executed as a strategy in a conversational exchange, as a misleading or deceptive tactic to get the best of one's speech partner illicitly, which makes the argument properly subject to strong refutation, if the charge that the argument is a fallacy is sustained.

6. When Is a Fallacy Really a Fallacy?

The approach of the Amsterdam School is a very good one, because it does take the context of use of an argument into account. And as acknowledged in Walton (1989a, chap. 1), a key element in understanding why the fallacies are incorrect arguments is that they go against various rules of argumentation appropriate for conducting a critical discussion. Unfortunately, however, there is no one-to-one correspondence between the individual fallacies and violations of the rules of a critical discussion. The process of identifying and analyzing the fallacies must go deeper than just equating a fallacy with a violation of a rule of a critical discussion and for several reasons.

One reason is that the types of argumentation identified with the various fallacies are not always fallacious arguments. For example, the ad hominem argument, although traditionally classified as a fallacy, is, in some instances, a reasonable (nonfallacious) argument. Or at least it will be a contention of this book, supported by other recent research on the fallacies, that many of the so-called fallacies are, in specific instances, not used as fallacious arguments.[8] As a result it is more difficult to analyze the fallacies than has previously been thought. Van Eemeren and Grootendorst (1984; 1987; 1992), however, presume that the so-called fallacies (or the arguments identified with them) are types of argumentation that can and should be defined as inherently fallacious. For example, they define the *ad verecundiam*, *ad populum*, and other types of arguments of this sort, as fallacious.[9]

A more careful examination of the data furnished by case studies of these types of arguments, however, reveals that they fall into three categories of evaluation: (1) those that are reasonable arguments, (2) those that are weak or inadequately supported arguments, and (3) those that are fallacious.[10] The problem with the view of van Eemeren and Grootendorst is that it sees all violations of the rules of a critical discussion as fallacious. This procedure fails to distinguish between the relatively trivial violations—blunders (nonfallacious errors that are failures to support an argument adequately)—and fallacies (more serious, systematic, underlying errors, or deceptive tactics used), which mean that an argument is radically wrong, from a logical point of view, in a way that makes it more difficult (or perhaps even impossible) to repair.

Unfortunately, you can't distinguish one fallacy (as a type of sophistical argumentation) from another by virtue of its violation of one rule of a critical discussion as opposed to another. Many of the fallacies, for example, are violations of the rule that requires an argument to be relevant in a critical discussion or the rule that requires an argument to fulfill a burden of proof. Such a violation does not identify the failure as a distinctive fallacy of this or that type (or even as a fallacy at all).

As we will see in this book, the fallacies are first and foremost identified as being certain distinctive types of arguments, as indicated by being instances of their characteristic argumentation schemes. These characteristic argumentation schemes are identified in chapter 5. Each has a distinctive form as a type of argument.[11] The ad hominem argument can be clearly distinguished, for example, from the ad verecundiam argument using these schemes. Then the fallacy is analyzed as a certain type of misuse of the argumentation scheme.

This way of proceeding leads us to the evaluation problem. Many of the traditional so-called fallacies have often been described in the logic texts in a superficial way that would make them better classified as pseudofallacies. This practice relates to the general problem of fallacy names, discussed in chapter 7, section 4. For example, although the argumentum ad verecundiam is often described as the argument from authority or appeal to authority, the appeal to expert opinion is primarily meant. Yet with the advent of expert systems, it has become clear that the use of expert opinion in argument, if carried out properly, is not fallacious (per se). It is a weak (presumptive) kind of argumentation, but it can be a legitimate and correct kind of reasoned argumentation in many cases.

The argumentum ad verecundiam literally means "the argument from (or to) respect, reverence, or modesty." Locke invented this phrase—see chapter 6 and Hamblin (1970, 160)—to refer to the use of an opinion of a reputation expert to browbeat an adversary in argument by suggesting that this adversary would be thought to have committed a "breach of respect" to question the authority of such a dignified expert. Thus the fallacy here turns out (chap. 9, sec. 2) to be not the appeal to expertise, or even to authority per se, as a kind of argument but the misuse of this type of argument as a technique for browbeating an opponent in a dialogue in an effort to make him unable to carry on effectively with reasoned discussion any further. It is a sophistical tactic that can be used to subvert or seal off reasoned dialogue by trying to push an appeal to expert opinion too aggressively, making it appear to be something it is not. Only when so misused is such an argument correctly said to be fallacious.

Part of the problem is a linguistic shift in the meaning of the term 'fallacy' itself. Many arguments, like appeal to expert opinion, have through the evolution of a tradition become labeled as "fallacies." But they are, in principle, as we have so often seen, reasonable arguments (that only go wrong in some cases). Does this mean a "fallacy" can sometimes be a perfectly reasonable argument? It should not. But paradoxically, the tradition suggests this interpretation, confusing readers of the logic textbooks on how they should interpret the word 'fallacy.'

According to the new theory, a fallacy is (first and foremost) an argumentation scheme used wrongly. In the various chapters, we will see how the types of arguments corresponding to the traditional so-called fallacies have underlying argumentation schemes. If an argument of one of these types is advanced in the format of the appropriate type of dialogue and is backed up sufficiently in that context by the support of its distinctive premises, it can be a reasonable argument as used in that context of dialogue. To say that such an argument is reasonable, however, is not generally to say only that it has a certain structure of constants and variables in its premises and conclusion similar to that which one finds in a deductively valid argument. Instead, it is to say that the argument is a sequence of argumentation that contributes to the realization of a proper goal of dialogue for the context in which it was advanced.

Each argumentation scheme has a matching set of critical questions, to be advanced by the respondent in the dialogue. To raise a critical question is to shift a burden of proof back onto the proponent who advanced the particular argument scheme in the first place. But to raise such a question is not to accuse the proponent of committing a fallacy or to claim that his argument is fallacious.

The idea of fallacy arises through the possibility that argumentation schemes and themes can be used wrongly, as calculated mechanisms of preventing appropriate critical questions from arising at all, by impeding the dialogue in certain characteristic ways. This new concept of a fallacy is premised on argumentation schemes that are inherently presumptive in nature, that is, that come into play as arguments where knowledge is insufficient to derive a conclusion with certainty or even with probability. Such cases of balance-of-considerations argumentation are settled on the basis of burden of proof in a dialogue. Traditionally, however, the reputation of this type of argumentation for subjectivity has led mainstream logic to be very suspicious of it as a respectable kind of reasoning at all (see sec. 2, above). But this suspicion must be dealt with, and overcome, if we are to have a logical theory useful for identifying, analyzing, and evaluating fallacies.

7. Persuasion Dialogue

In this book, the critical discussion, of the type identified by van Eemeren and Grootendorst, will be classified as a subspecies of a more general type of dialogue called persuasion dialogue. In a *persua-*

sion dialogue (Walton 1989a, 6) both parties have a *thesis*, a proposition to be proved to the other party, based on premises that are commitments of that other party. The goal of each party is essentially to persuade the other party that something is true.

But what is the goal of the persuasion dialogue as a collaborative social activity, freely entered into by both parties? According to van Eemeren and Grootendorst, the goal of a critical discussion is to resolve a conflict of opinions. But it seems that persuasion dialogues are often quite instructive and successful even if the original conflict of opinions was not "resolved," in the sense that one side's opinion was proved true and the other side's was proved false. Indeed, this very lack of a decisive resolution is often the basis of many critics' reservations about persuasion dialogue as a way of getting at the truth of a matter.

One might object that dialogue-based reasoning can generally never yield truth, or objective proof of a matter, because it yields only conclusions based on presumptions that could themselves turn out to be erroneous. Witness the horrifying fallacies that persuasion dialogue is susceptible to. The dangers, pitfalls, and fallacies of dialectical reasoning—and its inherently questioning nature as a kind of presumptive reasoning to be contrasted with the more cumulative methods of inquiry professed by the natural sciences—are lessons that have been abundantly and emphatically revealed by the study of fallacies.

Even if a persuasion dialogue does not generate a conclusion that is known to be true, still, the argument could be a good one by some other standard. Many arguments, it seems, are practically useful even if the premises and conclusion are not known to be true or are not established beyond doubt. Critical discussion as a type of argumentation does not aim at absolutely establishing a conclusion on the basis of what Locke called the *argumentum ad judicium*, or "proofs drawn from any of the foundations of knowledge or probability." A critical discussion has the more limited aim of digging into the reasoning behind a claim or commitment to see whether it is justified as a plausible presumption or not in relation to the opposed points of view of the two participants engaged in the persuasion dialogue. The aim is not (absolute) truth or probability but only tentative commitment as based on critical discussion of an issue. But here the skeptic will still be skeptical and will claim that argued commitment is not worth having—it is too subjective and unreliable—and is therefore not trustworthy as a source of scientific evidence.

The reply to this objection is once again to maintain that presumptive reasoning in persuasion dialogue is not meant as a substitute for scientific evidence drawn "from the nature of things themselves." As

Locke (see chap. 9, sec. 2) very astutely put it, only the argumentum ad judicium "brings true instruction with it and advances us in our way to knowledge." The function of critical persuasion in reasoned dialogue is, in Locke's terms again, to "dispose" us "for the reception of truth" by revealing our biases, prejudices, errors, blunders, and fallacies. By exposing the critical weaknesses in argumentation, persuasion dialogue can strengthen, deepen, and clarify an arguer's position, making it more worthy of acceptance if good reasons can be given that reply adequately to the right critical questions.

According to van Eemeren and Grootendorst (1984, 75), the goal of the critical discussion is to "resolve a dispute about an expressed opinion." The approach put forward in Walton (1984; 1987) stresses, however, that a persuasion dialogue can be very valuable and instructive even if the conflict is not resolved. This value arises through the maieutic function of dialogue, which allows a participant to gain an increased understanding or insight into the basis of his (or the other party's) position as the dialogue progresses. This statement is not to deny that the ostensible or explicit goal of a critical discussion is to resolve a conflict of opinions. But it is to add that a persuasion dialogue can often be successful in an important way, even if it is not completely successful in resolving the conflict of opinions. Even if the conflict is not resolved, the resulting maieutic insight gained in a dialogue can be a valuable benefit.

The gain in such cases is not empirical knowledge, or a resolution of the conflict, but rather a kind of self-knowledge or personal insight that can prepare the way for knowledge, or for the resolution of the conflict at some later point, by destroying a participant's prejudices, biases, or fallacies. The critical discussion, according to this view of it, serves as a mechanism to test one's personal commitments on an issue against the objections of someone who has a clearly defined position on the other side of the issue and who can argue for his position in an intelligent and forceful way. Just as a scientific hypothesis must stand the test of empirical data to become stronger and more carefully formulated, so must a personal commitment on a disputable issue stand the test of a critical discussion. By seeing the strongest arguments against your own point of view, you can not only strengthen that point of view and the position it is based on but also make it "deeper." A critical discussion that has this maieutic effect should be considered successful to some extent and beneficial as an exercise of argumentation for the participant and the community of discussants he interacts with.

So conceived, critical discussion, as a species of persuasion dialogue, can be justified as a reasoned method of argumentation that leads to reasoned commitment or new insight when it is properly executed. In critical discussion, the strongest arguments of each side are (or should be) tested out against each other. This represents the "experimental test" of a given presumption in the fires of contested argument. Subject to criticisms of the sort investigated in the next two chapters, an argument can be either refuted or refined and deepened, strengthened through the tests of hostile criticism and attack. In the process a successful argument becomes stronger as an argument through its use of an instrument by its proponent to defend his point of view. It is not by this positive means that dialectical reasoning in persuasion dialogue elicits new knowledge and understanding, however. For it is a notorious fact, strongly evidenced by the cases studied in this book, that a powerful and effective argument used to prevail on the assent of a respondent in dialectical reasoning can turn out to be fallacious or erroneous.

The real gain of knowledge from opinion-based reasoning in a critical discussion comes through the back door. Because the respondent must strive to deepen and strengthen her position (or abandon it) against the test of powerful criticisms, curiously she is led to a deepened understanding of her opponent's point of view. This is the maieutic function of dialogue—its capability to add to self-knowledge by deepening one's own understanding of one's deeply held convictions by revealing the reasons behind them. The maieutic function of well-executed dialogue is to take away the veil of ignorance that darkens one's most passionately held commitments, by exposing them to the light of criticisms and analysis. The result is a kind of destruction of ignorance that opens the way to knowledge—it is not external knowledge that comes from the "light arising from the nature of things themselves" in Locke's terms—but internal clearing away of the veil of dogmatism, bias, and fallacy that clarifies the basis of an arguer's internal convictions about the important things in life that are typically so subject to controversy and dispute.

At any rate, it is this concept of the maieutic function of persuasion dialogue that will be the chief rebuttal to the skeptical challenge to the theory of fallacy developed in this book, the challenge expressed by the charge "All dialogue-based argumentation is subjective and does not result in the truth being known." This skeptical point of view is hard to refute, but in this book a case is made against it.

According to the approach developed in this book, the internal clearing away of the veil of dogmatism, bias, and apparent correctness of bad arguments in dialogue is accomplished by the concept of fallacy. The concept of fallacy allows us justifiably and correctly (relative to a normative model of dialogue) to judge certain particular arguments as definitely incorrect. Our standards of correctness must be realistic, however, in relation to goals appropriate for argumentation in everyday conversations.

According to this approach, to judge the success of a critical discussion exclusively on the basis of the resolution of the conflict of opinions in the dialogue can be seen as a somewhat narrow approach with some important classes of cases. Hence we will argue that many cases of everyday argumentation can more realistically be viewed as instances of persuasion dialogue that is successful even if the conflict of opinions is not resolved.

Typically persuasion dialogues are on controversial questions of ethics, public policies, and the like that are inherently opinion-based. The reasoning in such cases is a kind of presumptive or default argumentation that does not conclusively resolve or settle the issue. Even so, such discussions can be insightful and informative, with a genuine educational value that enables the participants to deal with the conflict in a more tolerant and constructive way that may make them better able to appreciate the subtleties involved and the depth of the issue.

Demanding a resolution of the conflict in order to judge a critical discussion as successful would seem to imply that a "true" or "right" answer is always possible on the question that is the issue of such a discussion. But the problem is that persuasion dialogue does not generally result in an outcome that can be said to be known to be true or false. Critical discussion is particularly useful and appropriate precisely in instances where in fact that is not the case. In this type of situation, the outcome of the discussion, even though it may be partly based on knowledge introduced into the dialogue, is also partly based on presumptions that are inherently subject to controversy at the present stage of knowledge. In such a case, a critical discussion type of dialogue (persuasion dialogue) can be informative and useful, even if it reaches only a tentative conclusion based on presumption and burden of proof.

8. Profiles of Dialogue

Fallacies are committed where an argumentation scheme is used at a particular point in a dialogue, in a manner that fouls up the right sequence of questions and answers appropriate for that particular stage of the dialogue. Thus the concept of fallacy is associated not just with an incorrect or insufficiently supported argumentation scheme or with a violation of a rule of the dialogue. Instead, the sequence of moves and countermoves (the argumentation theme or technique) reveals a tactic that is used to block or subvert the goal of the dialogue. To see this point, and to identify the fallacy, you need to map the sequence of moves in the given case, as reconstructed from the text of discourse, onto the normative model of the sequence that should (ideally) have taken place.

For this purpose Krabbe (1992, 277–81) recommends the method of profiles of dialogue, used in Walton (1989a, 68–69; 1989b, 37–38), to analyze and evaluate cases of the fallacy of many questions, like "Have you stopped cheating on your income tax?" According to Krabbe (277), "Profiles of dialogue are tree-shaped descriptions of sequences of dialectic moves that display the various ways a reasonable dialogue could proceed." Krabbe points out that the method of profiles is useful because it enables one to discuss fallacies and other critical moves in a dialogue "without having to go through all the technical preliminaries necessary for the complete definition of a dialogue system" (ibid.).

A simple, illustrative example of a type of profile to begin discussion of the income tax question above might be the one in figure 1. Figure 1 already indicates something about the tactic used in this type of question. It is a yes-no question that admits of only two direct answers, but whichever answer is given, the respondent concedes having done something very bad, something that might tend to destroy her credibility in the subsequent persuasion dialogue. More extensive profiles used to analyze this fallacy are given in chapter 7, section 2.[12]

The important thing about a profile is that it is more than a simple adjacency pair or one move paired with a next move. And it is not just a localized argument with several premises and a conclusion of the kind we are so familiar with in logic. It is a sequence of connected moves that makes sense, or illustrates a familiar routine in everyday conversation, in light of some conventionalized type of dialogue that we generally understand as a verbal interaction between two speakers. The profile reveals a kind of tactic that is characteristically used to try to get the best of a speech partner in dialogue unfairly or deceptively. In this instance, it is the use of a loaded question in an inappropriately aggressive way to try to force the respondent to concede guilt.

The inherent nature of fallacy, according to the theory given in chapter 8, is to be found in the Gricean principle of cooperativeness, which says that you must make the kind of contribution required to move a dialogue forward at that specific stage of the dialogue. This principle requires, at any given point in a dialogue, a certain kind of sequence of moves to make the dialogue go forward. Each participant has to take proper turns, first of all. Then once one participant has

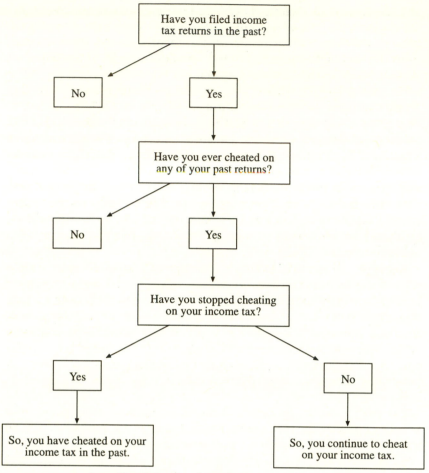

Figure 1 Initial Profile of the Tax Question

made a certain type of move, like asking a question, the other party must make a move that matches the previous move, like providing an appropriate response. A set of these matching moves and counter-moves is a connected sequence that makes up a profile of dialogue. This profile, when viewed in the proper context of dialogue, identifies the fallacy that occurred.

Fallacies come into a dialogue essentially because the profile gets balled up in a way that is obstructive. The one party tries to move ahead too fast, by making an important move that is not yet proper in the sequence. Or the one party tries to shut the other party up by closing off the dialogue prematurely or by shifting to a different type of dialogue. In such cases, the sequence may start out right, but then the moves start to happen in the wrong places in the sequence. Or key moves are left out of a sequence that should have been properly put in. The result is that the sequence is not in the right order required for that type of dialogue and at that particular stage of the dialogue. At this point a fallacy occurs, where the resulting disorder is a type of sequence that blocks the dialogue or impedes it seriously.

For example, in the case of a fallacious argumentum ad ignorantiam, the one party may put forth an assertion he has not proved, or has not even given any argumentation for at all, and then demand that the other party either accept it or disprove it. Here each individual move in the dialogue is all right, but what has gone wrong is that the first party failed to give some support to his argument before making his move of demanding that the other party accept or disprove it. The fault here was the key missing move in the sequence (Walton 1992d; 1994a).

Of course, you could say that this case was simply a failure to fulfill the burden of proof, which is, of course, a violation of a rule of a critical discussion. But that failure, in itself, was not the fallacy. Mere failure to prove something is not itself a specific fallacy per se. What went wrong was the failure to do what was required at the right step in the sequence. Such a fallacy is only adequately modeled normatively by looking at the whole sequence of moves and seeing that one required move was missing. The profile of dialogue reveals the fallacy, not the single missing move by itself.

Another case in point is the fallacy of begging the question. Again, the failure is one of an arguer trying to push ahead too aggressively in a dialogue by balling up the proper sequence. Instead of fulfilling burden of proof properly by an appropriate sequence of argumentation, the proponent tries to conceal this failure by pressing in a proposition that is in doubt as a premise (Walton 1991a). Once again, the fallacy is not simply the violation of fulfilling the requirement of burden of proof, although that is part of it. The fallacy can be identified only by looking at the whole sequence of argumentation, which could be done by using an argument diagram, or a profile of dialogue, and ascertaining that the sequence in the profile has a circular configuration. That is, the profile comes back to the same point or proposition previously in the sequence already. This profile must then be shown to be inappropriate for the given stage and context of dialogue. The actual profile must demonstrably fall short of the correct type of profile for that stage of a normative model of dialogue.

Of course, it is informative to say that such a sequence is wrong because a rule of a specific type of dialogue, like a critical discussion, has been broken. But that statement, in itself, is not sufficient to explain why a fallacy occurred or to determine which of the fallacies was committed. To do so, we must look at the profile of dialogue and see how the tactic that was used balled up the right sequence of dialogue in a particular way in order to identify the sophism.

With respect to some of the twenty-five major informal fallacies, in particular, where the fallacy is a paralogism, the order in the profile is determined by the kind of argumentation scheme that is appropriate, the accompanying critical questions matching that scheme. In these cases, identification of which fallacy has been committed can be carried out by identifying the argumentation scheme that was used. But this procedure works for only some cases of fallacies, not all. For example, if the argumentation scheme that was used was the negative argument from ethos, then if a fallacy occurred through the wrong use of this scheme in context, we can say that the fallacy that occurred was the abusive ad hominem.

But identifying or classifying a fallacy is different from evaluating or explaining a fallacy. If too little or no evidence was given to support the premise that a person has a bad character, would that make an ad hominem argument fallacious? Maybe not, if when asked to supply such evidence, the arguer complied or at least did not try to evade the request or show other evidence in the profile of dialogue of making inappropriate further moves.

Evaluating whether a particular case is fallacious or not, especially where the fallacy is a sophism, requires essential reference to the wider profile of dialogue. Knowing that the argumentation scheme was used incorrectly is not, in itself, sufficient for such a determination. The reason is that if an argumentation scheme was used incorrectly, it could have been a slip or an oversight. Much may depend on the kind of follow-up moves made in response to the other party's critical questioning of the move. It does not follow, in every instance, that a fallacy was committed. The reason is that there is a difference between an error in argumentation and a fallacy. A fallacy is a particularly serious kind of error, or infraction of the rules of dialogue, identified with a baptizable type of argumentation that has been abused in such a way as to impede the goals of the type of dialogue the participants in the argumentation were rightly supposed to be engaged in.

9. Argumentation Tactics

Informal logic and the study of fallacies generally involve the correct use of argumentation schemes and therefore parallel formal logic, which is based on forms of argument, like *modus ponens* and so forth. But informal logic also has to do with *argument strategies*, that is, with sequences of argumentation used to move toward goals of dialogue. *Argument tactics* are more locally specific pieces of advice that tell a participant in argument how best to achieve goals of dialogue in a specific situation. Tactical advice rules tell a participant how to make moves in argumentation that will help her play her part in the game more effectively. More specifically, tactics are most appropriate when the game has an adversarial element. In that type of context, tactics are devices that enable a participant in argumentation to defend her position more effectively or to attack or challenge the arguments of her opponents more effectively.

Tactics in argumentation can be codified as rules. As we will see in chapter 4, however, it is important to distinguish carefully between tactical rules and win-loss rules in a game of dialogue. The win-loss rules define the sequences of moves that constitute a win or loss of the game. For example, in a persuasion dialogue, a player is said to win the game (achieve his goal, fulfill his burden of proof) if he proves his own thesis as a conclusion, using only premises that his opponent is committed to along with the rules of inference allowed in the game. Thus the win-loss rules of a game define specifically what counts as a winning sequence of play, or a losing sequence of play. In effect, the win-loss rules define the goal of argumentation in a context of dialogue.

Tactical rules are different, however. They are more localized and more tailored to particular situations that arise in junctures of play during certain types of points in the sequence of a game. Tactical rules of argumentation are like coaching strategies that can be used in any kind of competitive game to train a student of the art in question to react effectively to types of moves that will be made by an opponent. Tactical rules are tips that help a player to attack and defend more effectively during a critical juncture in the game. They don't define what is a win or a loss, but they help you achieve a win in tricky situations where you could easily lose to your opponent's move.

Flowers, McGuire, and Birnbaum (1982, 280) define *argument tactics* as rules that describe the options on how to go about attacking or defending a proposition based on argument relationships that support or challenge that proposition. They define argument tactics in terms of argument relationships. One kind of argument relationship is that of *support,* where one point is evidence for another point (279). Another kind of argument relationship is that of *attack,* where one point challenges another point (279). This is a good kind of definition, because it divides tactics into the two basic types—attacking and defending tactics. It is an appropriate definition for the context of dialogue that Flowers, McGuire, and Birnbaum are mainly concerned with in their paper (1982), namely what they call *adversary arguments*, where neither participant expects to be persuaded and where both participants intend to remain adversaries in presenting their arguments to an audience.

The concept of argument tactics needs to be included within the broader category of argumentation techniques, however, so that it can apply to many different contexts of dialogue, including both adversarial and collaborative contexts. While argument tactics are clearly very important in understanding adversarial argumentation in cases of the argumentum ad hominem, argumentation techniques, more broadly conceived, can be equally important in helping us to understand arguments like the ad verecundiam cases that occur in a context of expert-layperson advice giving. Such contexts of dialogue may be basically collaborative rather than adversarial. But even so, techniques for cross-examining the expert in order to solicit clear, useful, and relevant advice may be very important. Thus although the military connotations of the term 'tactics' may suggest outright adversarial warfare, the concept of a technique of argumentation should also cover cases where the goal of dialogue is not primarily or exclusively to defeat or attack the other party in order to win the exchange.

In a broad sense, then, argumentation tactics and techniques can be codified in rules or heuristic pieces of advice that counsel a participant on how to fulfill his goal in a particular context of dialogue in certain characteristic types of situations that are likely to arise in that kind of dialogue. Such tactical rules advise a participant on how to defend his own arguments in the exchange and how to criticize or attack the arguments advanced by the other participant. Argument tactics are always related to, and determined by, the appropriate argumentation scheme at a particular point in a sequence of dialogue, and are related to the critical questions appropriate for that scheme.

Tactics are closely related to, and are localized parts of, strategies in dialogue. Strategies are more global and more general long-term types of sequences of moves toward a goal in dialogue. Tactics tend to be more localized parts of strategies that function as substrategies tailored to the specifics of a particular situation that has developed at some point in a dialogue. To glimpse how strategies and tactics work in games of dialogue, it is useful to look at an example of a formal game of dialogue in the literature. The formal game of dialogue CB was constructed in Walton (1984, 131–37) to model a case in a dispute called *Republic of Taronga*, where two foreign affairs specialists are having a discussion about economic developments in a fictional republic. The two specialists, Black and White, disagree about some of the relevant facts (premises), and they also disagree about a particular conclusion, taking opposed points of view on it. To resolve this conflict of opinion, they argue.

The game of dialogue arises because Black and White each argue by trying to get the other to accept premises (commitments) that will logically imply her own thesis (point of view) by the deductive rules of inference allowed by the game. The *problem of strategy* is posed by the fact that each player realizes that the other player will not commit to a proposition that directly or obviously implies the first player's thesis according to the rules of inference. For to make such a commitment would result, very quickly, in losing the game. Therefore, each player must devise strategies to break up the required commitments into smaller parts or must otherwise attempt to conceal their real import as proofs, so that the other player is less likely to be disinclined to accept them when asked to concede them as premises. Schopenhauer (1951) recognized this argument strategy precisely in formulating his ninth stratagem of controversial dialectic (21):

> If you want to draw a conclusion, you must not let it be foreseen, but you must get the premises admitted one by one, unobserved, mingling them here and there in your talk; otherwise, your opponent will attempt all sorts of chicanery.

This rule states a general strategy of argumentation. But as applied to a particular case, it can also be seen as a tactical rule that offers an arguer practical advice on how to get the best of an opponent.

Studying this strategic problem of reasoned persuasion led to the formulation of a formal game of dialogue called CB (Walton 1984, 131). In addition to locution rules, commitment rules and dialogue rules, CB had what were called *strategic rules*, which combined win-loss rules with a rule that awarded points in the game as a kind of incentive for accepting premises. This incentive was a way of attempting to overcome a major problem with this type of game—a player might tend never to accept new premises when asked, always replying 'No commitment.' In retrospect, however, it is possible to see that the device of offering points to accept commitments in CB was an ad hoc solution.

Even so, CB was an interesting experiment in the development of formal games of dialogue, because some important kinds of strategies could be formulated in the game. For example, the *distancing strategy* is to ask your opponent to concede a proposition that is only distantly related to the thesis at issue and then fill in the gaps needed to deduce your thesis from it (Walton 1984, 142). Schopenhauer (1951, 21) also recognized this strategy in the latter part of his ninth stratagem where he suggests: "If it is doubtful whether your opponent will admit [premises], you must advance the premisses of these premisses." Other devices, called *strategy sets* (Walton 1984, 152), involve assessments of how deeply committed your opponent is to different members of a set of commitments. It was shown how depth of commitment was an important factor in strategy of argumentation in CB.

In order to model the argumentum ad hominem, a different approach to the problem of inducing a player to accept commitments was proposed in a game SBZ, a modification of a game CBZ that is an extension of the game CB. In CBZ and SBZ a new type of rule, called a dark-side rule, was proposed. The idea behind this rule was that each player's commitment set should be divided into two sides, a light side and a dark side. The light side contains all the commitments that a player realizes explicitly are commitments of his. The dark side contains propositions that are commitments of the player, but he does not know, or fully realize, that these propositions are commitments of his.

How strategy works in CBZ is strongly influenced by these dark-side commitments because of the following rule (the dark-side rule, RDS).

(RDS) If a player states 'No commitment A,' for any proposition A, and A is in the dark side of his commitment store, then A is immediately transferred to the light side of his commitment store.

The strategy for CBZ, then, is the following. If a player wants to get his opponent to accept a set of premises, he leaves gaps only where the propositions are dark-side commitments of his opponent. Then, at the last minute, he can fill in these gaps using the rule (RDS). This is effective strategy because the opponent cannot clearly see in advance what is going to happen, for he is unaware of his dark-side commitments.

The whole idea of the dark-side commitment sets is based on the Socratic philosophy that when we are in dialogue, reasoning with a questioner, we can come to see a participant's deeply held but unarticulated convictions more clearly. This is the *maieutic function* of dialogue, where the elenchic questioner can assist, like a midwife, in the birth of a new idea. The bringing of the new idea from the dark to the light of explicit commitment represents the birth of a new insight. It is also represented by Plato's myth of the cave—making a dark or murky commitment become clear represents the ascent from the cave to the light. Through questioning, the participant in dialogue is led to self-knowledge by a clarification of his own previously held (but dark) commitments.[13]

The most refined versions of the games with dark-side commitments are the games ABV and CBV in Walton (1987, 125–31). These games have the same kind of rule represented by (RDS) above. The V stands for 'veiled commitment-set,' where part of a commitment set is a dark side. Walton (1987, chap. 5) shows how dialogue games like CBV that have veiled commitment sets can be used to study the problem of unexpressed premises in argumentation. Strategy in games of dialogue is shown to be the key to this problem.

A recent research project, undertaken jointly by the author and Erik Krabbe, studies dialectical shifts from a looser and more friendly context of dialogue to a tightened-up context where commitment is indicated only by explicit concessions or retractions in explicit speech acts in a dialogue. The forthcoming monograph *Commitment in Dialogue* uses a CBV type of dialogue to model the first context and a more strictly formulated game of formal dialectic, of the type found in Krabbe (1985), to model the second context of dialogue. The results of this research indicate that strategic rules can be altered radically when there is a shift from looser to tighter standards of commitment.

10. Standards of Evaluation

To evaluate argumentation as being correct or not, and thereby to evaluate an argument in a given case as fallacious or not, one must understand the goals of the dialogue and also apply a normative model of dialogue as indicating a standard of correct use. One must then look at the particulars of the text of discourse in the given case, interpreting the given argument from the text. Of course, in many cases, not enough information is given to enable us to judge what the argument is. In such cases, we can, at best, conditionally evaluate whether the argument is fallacious or not. Indeed, it is typical of the short examples usually given in the standard treatment of the textbooks that not enough context is provided to determine whether the argument (or the part given) is fallacious or not. We have to be prepared for this kind of problem, however, when attempting to do informal (applied) logic.

The reader has now been prepared for what is to come, namely that in this book we will be judging arguments in relation to their use in a context of dialogue (as far as this is known in the given case). Furthermore, these contexts of dialogue can change, so that the very same argument could be fallacious in one context but nonfallacious in another context. So conceived, fallaciousness will turn out to be a contextual matter of the type of conversation the arguer is supposedly engaged in.

No doubt this much contextualism (or relativism, you could call it) will be too much for many readers who were skeptical about making any sense of the fallacies in the first place. From this skeptical point of view, this very problem led generations of logicians to ignore or dismiss the fallacies—that they do not reduce to any absolute standard of validity or invalidity of arguments in the apparently context-free way that formal logic does. According to this point of view, our way of evaluating fallacies is too conditional on interpreting what was meant by an arguer, too dependent on the vagaries of natural language, too contextual, too subject to qualifications and potential exceptions, and, in a word, too subjective for logic. There is some truth in these objections, because the pragmatic theory given in this book aims at an applied art of judging arguments as used in a given case in a context of natural language conversation. Even so, it will be shown in detail how the skeptical point of view can be overcome.

This point of view, it will be argued, overlooks three vitally important factors. One factor is that each type of dialogue has rules and techniques of interactive argumentation that collectively define a normative model of (good) argumentation appropriate for a particular context of dialogue. These normative models are not empirical descriptions of dialogue behavior but analytical instruments that define sequences of argumentation that can be used rightly or wrongly, in an erroneous, blundering, or fallacious manner to violate the rules. When an argument is erroneous or fallacious, evidence can be given to back up the criticism of it by citing failures to meet requirements of a normative model in conjunction with textual evidence from the given discourse in a particular case.

A second factor is that we will give forms of argument, argumentation schemes, and themes for the various types of argumentation concerned, which can be used correctly in some instances and inappropriately in others. By judging a particular sequence of argumentative dialogue as a segment of discourse, it can be evaluated whether a particular argumentation scheme or theme has been used correctly or not. What you have to do is take the actual sequence of dialogue, as reconstructed from the text of discourse in the given case, and measure it up to the ideal sequence of discourse required by the normative model.

The third factor is that certain common and baptizable ways of arguing incorrectly and inappropriately in a dialogue, called fallacies, can be identified, analyzed, and evaluated so that we can learn to recognize them as incorrect arguments and deal with them when they occur. The reader will have to judge for herself whether a good enough case has been made out, in the book, for the conclusion that these three factors can be determined with enough of the right kind of evidence on which to base our new theory of fallacy.

Three problems confronted are that of fallacy identification, fallacy analysis, and fallacy evaluation. All three problems are solved by developing new pragmatic structures that display the form of an argument (the so-called argumentation scheme) and then show how this form fits into an enveloping normative structure of dialogue. In this book it is shown how the twenty-five or so major informal fallacies can be identified, analyzed, and evaluated with a promising degree of success using the structures set out. Each fallacy itself represents a nontrivial problem for analysis, however.

It is shown in the book how the examples of the fallacies given in the logic textbooks characteristically turn out to be variants of arguments that are reasonable, even if defeasible or questionable, and are based on presumptive reasoning. This is the essence of the evaluation problem. It is a key thesis of the book that you must not take for granted, as the textbooks in the past have so often done, that you can spot a fallacy simply by looking at the type of argument it is, in abstraction from its use in a context of dialogue. This context is demonstrated to be especially important when dealing with the sophistical tactics type of fallacies, where the profile of dialogue is all-important in showing how the argument was used, for example, too aggressively, to bring undue pressure to bear on a participant.

The book argues that questionable arguments, and blunders in argumentation, need to be distinguished from fallacious arguments. It is stressed that the claim that an argument is fallacious should be seen as a strong form of condemnation that needs to be backed up by certain kinds of evidence that meet a burden of proof appropriate for such an allegation.

Formal logic has been successful as a scientific discipline because it is based on argument forms that can be evaluated as valid or invalid in an enveloping structure. What has been lacking, however, is an informal or practical logic that judges the use of an argument in a given case, in a context of conversation. The intent of this book is to contribute to a restoration of this imbalance by basing a theory of fallacy on argumentation schemes that can be evaluated on how they are used in different developing structures of dialogue.

Chapter 5 presents twenty-five of these basic argumentation schemes for presumptive reasoning, with sets of accompanying critical questions matching each scheme. The view of fallacy then presented is that a fallacy is either a paralogism, an argumentation scheme used in such a way that it systematically fails to answer a critical question appropriate for that scheme, or a sophism, a more extended misuse of a scheme or sequence of them connected together that has been twisted or used incorrectly in a dialogue, as evidenced by a distorted profile of dialogue, or what is called an argumentation theme. A theme is a sequence of connected moves in a dialogue, displayed by a profile, or tableau, a pair of matching columns of displayed moves. A profile showing evidence that a fallacy has been committed is one that has been distorted, or balled up. The moves are in an order or pattern that is normatively inappropriate in just this sense. The moves occur in a structurally blocking or interfering order, so that they do not forward the goal of the dialogue.

The paralogism type of fallacy is a systematic, underlying type of error of reasoning in an argument. The sophism type of fallacy occurs where a scheme is used as a deceptive tactic to try to get the best of the other party unfairly when two parties reason together in one or more of the several types of dialogue.

2

Informal Fallacies

This chapter is a survey of the twenty or so major informal fallacies that are typically featured in the standard treatment of fallacies in the logic textbooks. Most of these fallacies, and their names (translated in most cases from Greek to Latin), originated in Aristotle's manual on fallacies, *De sophisticis elenchis* (*On Sophistical Refutations*). At least four of them originated with Locke—see Hamblin (1970, 159–62)—and a few of them are of yet undetermined origin.

The descriptions of the fallacies, and the examples used to illustrate them, recur over and over again with different variations in the multitude of textbooks. Successive generations of textbooks seem to have taken pretty much the same material from the textbooks of the previous generations. Different texts have used different classifications and have often added small insights or novelties thought to be improvements on the tradition. But on the whole, things have not changed much in this field.

Hamblin (1970) described the standard treatment of fallacies as stale, superficial, and time-worn. It is an area where the old material has been passed on and taught, but where no serious research or scholarship has led to significant improvements or investigations of the logical structures that could be systematically used as a basis for identification, analysis, or evaluation of the fallacious arguments cited.

The textbooks have often been peppered with insights and good examples, however. The problem is not that the textbooks are bad or inherently wrong. The real problem is a scholarly one—the lack of some underlying theory.

The purpose of this chapter is not precisely identical to the first chapter of Hamblin (1970), which was to give a description of the standard treatment of the fallacies in the logic textbooks by taking a sample of several of the leading texts. Instead, the purpose here is to present the reader with a set of cases that graphically illustrate the type of wrong argument characteristic of the fallacy portrayed in the standard treatment. One problem with the standard treatment is that so many of the examples given are not clearly fallacious; many of them are reasonable arguments even if they are inconclusive or are open to critical questioning.

The cases in this chapter give us a rough set of reference points, at least some intuitive guidance on what the type of error is that each fallacy name is supposed to represent. Of course, not all texts agree, and there are certainly plenty of contradictions among them on how each fallacy is to be defined and understood as a type of error.

The problem is that for each fallacy cited in this chapter, as we will see later, there is a corresponding, similar type of argumentation that is nonfallacious. Hence it is important to begin with at least some relatively firm intuitive grasp of what is supposed to be fallacious about these common types of argumentation according to the traditions of the texts.

1. *Ad Hominem*

The argumentum ad hominem, or argument against the person (literally, "against the man"), is traditionally meant to denote the kind of argumentation that argues against somebody's argument by attacking the person who put forward the argument. Various types of argument against the person are recognized.

In the abusive ad hominem argument, the focus of the attack is the character of the person and, in particular, his character for veracity. According to Fearnside and Holther (1959, 99), personal attack is a common type of argument, "odious" yet effective: "There is no argument easier to construct or harder to combat than character assassination, and this may be the reason personal attacks are so commonly on the lips of ignorance and demagogy."

In the following case, Flora MacDonald, a member of the opposition party, questioned the prime minister of Canada in the oral question period of the House of Commons debates (House of Commons, 1984, 1457), on whether he had been keeping files on the leader of the opposition. Pierre Trudeau, then prime minister, replied that press clippings of public records are kept but that there were no investigations into private affairs to his knowledge. MacDonald replied:

Case 1

What the Prime Minister fails to realize is that every time an office of a Cabinet Minister sends out any kind of press release, it goes into a file in the Member's office. What he does not seem to understand is that there is a distinction between the Prime Minister of Canada using the power of the state to search a title of a private individual and that which is put into the files of any Member of Parliament in this House. That is the distinction. The problem is that the Prime Minister and his colleagues have been wallowing around in the sewers for so long that they no longer know the difference between right and wrong.

This argument is an abusive ad hominem because the character of the prime minister is attacked. It is also somewhat more general because the attack includes his colleagues as well. The argument is that we should not believe Trudeau's denial that investigations into private affairs were going on because of the bad moral character of Trudeau and his colleagues.

The circumstantial ad hominem argument is a questioning or criticizing of an arguer's position by citing a presumption of inconsistency in his position. Typically, the inconsistency alleged is a pragmatic (practical) inconsistency rather than a purely logical inconsistency, and the allegation often relates to personal actions or past conduct of the arguer criticized. The term 'circumstantial' is appropriate because the alleged inconsistency is between his personal circumstances and what he says in his argument. Hence the expression "You don't practice what you preach" characteristically expresses the thrust of this type of criticism.

The circumstantial type of ad hominem fallacy resides in a certain sort of comparison of cases or parallel. The classic case (Whately 1836, 196) is called the *Sportsman's Rejoinder*,[1] paraphrased below.

Case 2

A hunter accused of barbarity for his sacrifice of innocent animals for his own amusement or sport in hunting replies to his critic: "Why do you feed on the flesh of harmless cattle?"

Here the hunter tries to refute the critic by referring to the critic's own special circumstances (being a meat-eater). This case fits the argumentation for the circumstantial ad hominem argument, because the hunter is alleging that the critic is pragmatically inconsistent and that therefore her contention of barbarity should be rejected as not credible.

In the bias type of ad hominem argument, the attacker claims that the person attacked is not an impartial or credible exponent of the conclusion he advocates on the grounds that he is biased. The term 'bias' is a negative one, meaning that the person in question is so strongly committed to one side of an issue by partisan interests that he is not fairly taking the evidence on both sides into account.[2]

Case 3

Smith claims that higher taxes will inhibit the economy, contributing to the recession, but I wouldn't believe him on that issue. He and his conservative cronies always say that because they don't want taxes to reduce their profits in business. Smith owns a lot of stocks in many big businesses.

This argument need not involve an attack on Smith's character like that of the abusive ad hominem. It could involve such an attack as well. But the thrust of the bias ad hominem is somewhat different. What is under attack is Smith's ability to be a serious participant in this particular dialogue on the issue of higher taxes as an arguer who is really open to looking at evidence on both sides and to taking into account, or conceding, a good argument even if it goes against his side. A biased arguer won't admit fair defeat and will support his own side even if the evidence is against it.

The poisoning-the-well type of ad hominem argument is an extension of the bias type of ad hominem where the arguer is said to be so hopelessly biased, or "fixed" to one side, that nothing she says could ever be trusted as reliable, or taken at face value. Often this type of attack cites the arguer attacked as belonging to a particular group and therefore as representing the viewpoint of this group, without any possibility of her ever being able to escape this bias or to say anything not determined by it.

The following ad hominem imputation of irremediable bias occurred during a debate on abortion in the Canadian House of Commons (House of Commons 1979, 1920).

Case 4

I wish it were possible for men to get really emotionally involved in this question. It is really impossible for the man, for whom it is impossible to be in this situation, to really see it from the woman's point of view. That is why I am concerned that there are not more women in this House available to speak about this from the woman's point of view.

As noted in Walton (1989a, 51), this argument is based on a true assumption, namely that a man cannot personally experience an abortion. The implication, however, is that men are not qualified to speak on the issue because, as males, they are always subject to a bias that makes their viewpoint limited and deficient. No matter what a man says, then, it must always be discounted or rejected as biased. The poisoning-the-well type of ad hominem argument is particularly muzzling because no matter how good a person's argument is, it will always appear suspicious and unconvincing.

In some cases, one type of ad hominem argument is a kind of lead-in to the other. For example, it may be argued that Mr. *x* is circumstantially inconsistent and does not practice what he preaches. But this circumstantial attack may be followed up by arguing that Mr. *x* is therefore a hypocrite, an insincere type of person who does not truly mean what he says. This second part of the argumentation is an abusive species of ad hominem attack. Here the abusive arises out of the circumstantial ad hominem argument.

Case 5, a speech from the *Annals of the Congress of the United States* (November 2, 1812, to March 3, 1918, 540–70), cited by Brinton (1985, 56) and Walton (1989a, 170), included a bias ad hominem attack.

> *Case 5*
>
> The subject of debate in the U.S. Congress in 1813 was the New Army Bill, a proposal to raise more troops for the war against England. The majority, led by Speaker of the House Henry Clay, argued that an invasion of Canada with these additional troops would help to win the conflict. Josiah Quincy, speaking for the opposition on January 5, 1813, argued that the additional troops would be insufficient, that an invasion of Canada would be unsuccessful and immoral, that a conquest of Canada would not force England to negotiate, and finally that the bill was politically motivated, "as a means for the advancement of objects of personal or local ambition of the members of the American Cabinet."

Using a bias ad hominem attack, Quincy argued that his opponents were motivated by "personal or local ambition" and could not therefore be trusted as unbiased participants in the dialogue.

But then later in his speech, Quincy called his opponents "toads, or reptiles, which *spread their slime on the drawing room floor.*" This part of the argument was an abusive (direct) ad hominem attack.

There are two basic problems with the ad hominem fallacy. One is identifying it as a specific type of argument. For it seems that the abusive and circumstantial varieties really represent two distinct types of argument (not to mention the problem of the other varieties—see Krabbe and Walton 1993). The other problem is that these distinct types of argumentation appear to be reasonable (nonfallacious) in many instances. For example, attacking the character for veracity of a witness is an acceptable type of argumentation (within limits) in cross-examination in court. And accusing someone of "not practicing what he preaches" is, in principle, a legitimate way of criticizing the argument of someone who exhibits such a conflict (Walton 1992b, chap. 6).

Clearly much work remains to be done in defining the ad hominem as a clearly identifiable type of argument and in finding criteria to judge which instances of it are fallacious and which are not.

2. *Ad Baculum, ad Populum,* and *ad Misericordiam*

The textbooks define the *argumentum ad baculum* (argument to the club or stick) in three different but overlapping ways. It is said to be the use of an appeal to force, to fear, or to a threat, to cause acceptance of a conclusion.[3] An appeal to fear could include tactics of intimidation without a threat's being made. An appeal to force need not necessarily be an appeal to fear. Although force does seem to be connected to a threat, much depends on how you might define 'force' or 'threat.'

Quite often, the examples given by the textbooks involve covert threats (as opposed to overt threats), where it is said by a proponent to a respondent that bad consequences (i.e., consequences unfavorable to the respondent) will happen if the respondent does (or doesn't do) something. The following classic case of a covert threat is from Copi (1986, 106).

> *Case 6*
>
> According to R. Grunberger, author of *A Social History of the Third Reich,* published in Britain, the Nazis used to send the following notice to German readers who let their subscriptions lapse: "Our paper certainly deserves the support of every German. We shall continue to forward copies of it to you, and hope that you will not want to expose yourself to unfortunate consequences in the case of cancellation."

By contrast, a case of an overt threat is the sequence in the cartoon *Blondie* (King Features Syndicate, 1973), where a salesman comes to Dagwood's door (also in Walton 1992b, 164).

> *Case 7*
>
> **Salesman:** I'm selling this window cleaner. And I'm not a guy who likes to fool around. Either you buy it, or I'll punch your lights out!
> **Dagwood** (walking back into his living room after buying two bottles of window cleaner): He has a very persuasive sales approach.

Cases likely to deceive anyone seriously would probably be more subtle than these two cases. Of the two types, the covert ad baculum seems to have more potential for serious deceptions.

The first problem with the ad baculum is to identify it. Is it an appeal to threat, or more broadly, are scaremongering tactics that do not involve a threat counted? The second problem is that of evaluating it, for not all threats, or appeals to fear, are fallacious as arguments. For example, in union-management negotiations, a threat to take strike action can be a legitimate part of the argumentation in the bargaining process (Walton 1992b, 158).

The argumentum ad populum is the type of argument that appeals to popular sentiment to support a conclusion. It is sometimes also called "appeal to popular pieties," "appeal to the gallery," or, even more negatively, "mob appeal" by textbooks. According to Engel (1982, 173), such arguments are fallacious because they "steer us toward a conclusion by means of passion rather than reason," they "appeal to our lowest instincts," and they "invite people's unthinking acceptance of ideas which are presented in a strong, theatrical manner." What appears objectionable here is the emotional tone of a speech as a substitute for reason.

Other cases of ad populum could perhaps also be called the appeal-to-popularity argument, where an opinion is said to be universally held or universally held by a group whose opinion is held to be important. In the following case, Trevor and Grace are having a debate on capital punishment, and Grace argues:

Case 8

Every civilized country in the world has done away with capital punishment. People like you, who still believe in it, are out of the picture—Neanderthals!

Teenagers are adept at using this type of argumentation against parents when they say things like, "That's not how we do things now," suggesting that anyone who acts differently is out of the trendy mainstream and that therefore his argument can be discounted as worthless.

An inherent problem with the ad populum fallacy is that drawing conclusions on the basis of what one takes to be popularly accepted opinion, if properly qualified, can be a reasonable kind of argument, especially in a democratic system of politics (Walton 1992b, 69–90). Aristotle (*On Sophistical Refutations*, 165 b 3) even defined *dialectical argument* as a distinctive type of argument that reasons on the basis of premises that are generally accepted opinions. He did not classify this type of argumentation as a species of fallacy but saw it generally (possibly subject to exceptions) as a reasonable type of argument. How then can we distinguish between its reasonable and fallacious uses?

The argumentum ad misericordiam is the type of argument that uses an appeal to sympathy for human plight, compassion, or pity, to support a conclusion. A good example is given by Michalos (1970, 52).

> *Case 9*
>
> A student who missed practically every class and did nothing outside class to master the material told me that if he failed the course he would probably be drafted into the army.

It is easy to see that the student's argument in this case puts an inappropriate kind of pressure on the instructor, whose job is supposed to be to grade the student's work impartially, on the basis of its merit. According to Michalos (52), this case is a fallacy because the issue should be "not what happens if the student fails but whether or not he deserves to fail." The appeal is not relevant, despite its stimulating emotional appeal. Similar appeals of this type are familiar enough.

> *Case 10*
>
> If I don't get an *A* in this course, I won't get into law school, and my career plans will be ruined.

Another type of case is somewhat different from the previous two. When the decision was being made to commit American troops to liberate Kuwait, a fifteen-year-old Kuwaiti girl identified only as Nayirah (sobbing), testified before the Congressional Human Rights Caucus that she had seen Iraqi soldiers take babies out of incubators (*60 Minutes*, January 19, 1992).

> *Case 11*
>
> Nayirah testified that she saw the Iraqi soldiers come into the hospital with guns—"They took the babies out of the incubators, took the incubators, and left the children on the cold floor [crying]." The resolution to go to war passed in the U.S. Senate by only five votes. Seven senators mentioned the incubator atrocity in the debate on whether to go to war. Later, it was found that Nayirah was a member of the Kuwaiti royal family, and the daughter of Kuwait's ambassador to the U.S., but her true identity was only discovered later by an American reporter. Later inquiries could find no evidence that babies were pulled from incubators. It was found that the baby incubator story had been promoted by an American public relations firm with links to Kuwait.

Subsequent evidence in this case showed that the baby incubator story was a well-financed public relations tactic that was very successful in achieving its goal (Walton 1994b).

The ad baculum, ad populum, and ad misericordiam all appear to succeed because of the powerful impact of emotional appeals in argumentation, but it is shown in Walton (1992b) how these emotional appeals can often be quite reasonable arguments used to shift a burden of proof in a balance-of-considerations argument.

3. *Ad Ignorantiam*

The *argumentum ad ignorantiam*, or argument to ignorance, is said to take two forms. The first form occurs where it is concluded that a proposition is true on the basis that this proposition is not known (proved) to be false. The second form occurs where it is concluded that a proposition is false on the basis that it is not known (proved) to be true. The one form is a kind of opposite or negative form of the other, where the word 'true' is replaced by the word 'false' and vice versa. Both forms of argument are based on a premise of lack of knowledge (ignorance). Hence the rationale of the phrase argumentum ad ignorantiam as a name for this type of argument is clear.

The argumentum ad ignorantiam is generally held by the textbooks to be a fallacy. For example, Copi (1986, 94) describes it as a fallacy:

> Case 12
>
> The fallacy of *argumentum ad ignorantiam* is illustrated by the argument that there must be ghosts because no one has ever been able to prove that there aren't any. The *argumentum ad ignorantiam* is committed whenever it is argued that a proposition is true simply on the basis that it has not been proved false, or that it is false because it has not been proved true. But our ignorance of how to prove or disprove a proposition clearly does not establish either the truth or the falsehood of that proposition.

Copi's example of the ghosts argument certainly seems to involve some sort of bad or questionable step of reasoning, although perhaps the negated counterpart ad ignorantiam argument is less persuasive as an example of a clear-cut fallacy: it has never been proved that ghosts exist, therefore they don't exist. But also, much depends on how strongly the argument is expressed. If I conclude that it *must be false* that ghosts exist on the basis of this latter argument, I would seem to be making an error of leaping to too strong a conclusion on the basis of negative evidence or ignorance. If my conclusion is only that it is reasonable to presume that ghosts don't exist until some good evidence of their existence can be established, however, my argument begins to seem much more reasonable. This kind of argument may not be a fallacy at all.

Despite some of the rough edges of Copi's example, it is not too difficult to appreciate the sense of his warning about the argumentum ad ignorantiam as a potentially serious kind of error of reasoning in some instances. This is especially evident in the context of inductive or scientific reasoning about experimental confirmation of a hypothesis. Absence of experimental support for a hypothesis is different from an experimental result that refutes, falsifies, or goes against the hypothesis. Because we do not yet have any data relevant to a hypothesis, it does not follow that the hypothesis should be rejected. Lack of confirmation does not necessarily imply disconfirmation of a hypothesis.

Another case in point would be mathematical reasoning. While it may be true that a certain mathematical conjecture has never been proved, it does not necessarily follow that it can't be proved. For it may be that the proposition is very difficult to prove and that nobody has succeeded in proving it yet. To show that a proposition cannot be proved makes an impossibility claim making it necessary to do more than simply cite ignorance of how to prove or indicate previous failures to prove it by mathematicians who worked very hard.

One type of case where the argument from ignorance is used as quite a seriously mischievous tactic of argumentation concerns the bringing forward of damaging charges made purely on the basis of innuendo and suspicion. The following case was used as an exercise by Copi and Cohen (1990, 107–8) to illustrate a fallacious argument from ignorance.

Case 13

On the Senate floor in 1950, Joe McCarthy announced that he had penetrated "Truman's iron curtain of secrecy." He had 81 case histories of persons whom he considered to be Communists in the State Department. Of Case 40, he said, "I do not have much information on this except the general statement of the agency that there is nothing in the files to disprove his Communist connections." [Rovere, 1959, 132]

In this case, we readily accept the evaluation of the argument from ignorance as fallacious because we know as a historical fact that the McCarthy investigation was a kind of "witch hunt" that used unfair methods to attack political enemies (or persons so perceived) by labeling them as Communists or "Communist sympathizers." As case 13 illustrates, once the frenzy mounted, even the absence of evidence to disprove such a charge was taken as a license to attack a victim by labeling such a person as a Communist. Many people lost their jobs because of these allegations.

Many of the individuals charged may have been Communists (not that the affiliation was a good reason for dismissing them). But the ad ignorantiam problem arose because they were condemned on the basis of suspicion or gossip or simply because someone who didn't like them pointed the finger of suspicion at them. Because of an innuendo or "smear" effect, such a charge may be such a nasty allegation, or may be so perceived, that it damages a person's reputation, leaving her under a cloud of suspicion. Yet the charge may have been based on no real evidence, just ignorance or an absence of evidence to refute the charge. Unfortunately, once such a colorful and personally damaging charge has been made, it may be very difficult to refute it even if there was no evidence to support it in the first place.

Another type of case where the argumentum ad ignorantiam strongly appears to be used fallaciously as a sophistical tactic is the citing of lack of evidence (falsely) as a "stonewalling" argument. In the following case, the U.S. Food and Drug Administration had known about research questioning the safety of polyurethane foam breast implants. The FDA and the manufacturers of the implants, however, tried to allay fears by not acknowledging the troubling findings on the grounds that they were "anecdotal" only, that is, that they were based only on complaints made by women rather than on "scientific" data.

Case 14

The FDA has known for months about the research questioning the safety of the foam implants, but the agency didn't acknowledge the troubling findings until last week. . . . Bristol-Myers Squibb also tried to allay the fears of women who have the implants. "Medical literature contains no reported cases of human cancer associated with polyurethane foam," said a company statement. But many women are worried. Sybil Goldrich and Kathleen Anneken, founders of Command Trust Network, a national information and support group for women with implants, report that their 24-hour hot line has been flooded with hundreds of calls since last week. [Seligmann, Yoffe, and Hager 1991, 56]

In this case, the claim that the medical literature contained no reported cases of human cancer associated with polyurethane breast implants may have been true but may conceal knowledge of plenty of disturbing cases (unofficially) reported by women—cases that could be good practical grounds indicating grave reservations for women considering having these implants.

In short then, we can see why the argumentum ad ignorantiam has been thought to be a fallacy. Because a proposition has not been proved, it does not necessarily follow that it can't be proved and therefore that it must be false or should be disregarded altogether. Difficulty of proof and confirmation of disproof are two separate things.

It has become more and more widely recognized by the textbooks, however, that arguments from ignorance can be reasonable in some cases. Many of the textbooks cite as an example the principle of criminal law that one is presumed to be not guilty until proven otherwise. This is a form of argumentation from ignorance, but it is definitely not a fallacious argument. Many other examples of nonfallacious arguments from ignorance have been cited in Walton (1992d)—for some examples of this sort, see cases 83–85, below. Hence it seems that the problem with arguments from ignorance is to determine when they are fallacious and when not. The cases above, however, show at least that this type of argumentation can be fallacious in some instances.

4. *Ad Verecundiam*

The expression *argumentum ad verecundiam* means appeal to reverence (respect) and refers to the fallacy of inappropriate use of appeals to expert opinion in argumentation. Despite this apparently peculiar phrase however, it is clear from the textbooks that the fallacy referred to is inappropriate appeal to authority, especially the authority of expert opinion. But when is such an appeal used inappropriately?

According to Copi and Cohen (1990, 95), "the fallacy of *ad verecundiam* arises when the appeal is made to parties having no legitimate claim to authority in the matter at hand." They cite the following example (95).

Case 15

In an argument about morality, an appeal to the opinions of Darwin, a towering authority in biology, would be fallacious.

Copi and Cohen add a qualification, however (95): "If the role of biology in moral questions were in dispute, Darwin might indeed be an appropriate authority." The fallacy occurs because Darwin is not an appropriate source to cite as an appropriate expert opinion on the subject of morality. The field of expertise is wrong.

It is not hard to imagine cases, however, where an ad verecundiam type of fallacy could occur even if the party appealed to does have a legitimate claim to being an expert, with qualifications. Consider a case where an advocate of a sugar-free diet, Paula, brought in a researcher and author who has a Ph.D. in nutrition to give a lecture on diet. The lecturer claimed that people shouldn't eat sugar because it causes food allergies, endocrine problems, hypoglycemia, diabetes, tooth decay, gum disease, osteoporosis, heart disease, arthritis, and cancer. A person from the audience, Herbert, made some objections during the question period that followed the talk. But Paula intervened, supporting the viewpoint of the lecturer.

Case 16

Herbert: I think you need sugar to stay alive, and anyway if you ate no sugar at all, your body would make glucose (a form of sugar) anyway, out of whatever you did eat. I can't believe that sugar causes all those disorders.
Paula: Well, what do you know about it anyway? Are you a nutritionist?
Herbert: No.
Paula: Just what I thought. Next question.

In this type of case, it would seem to be appropriate to say that Paula committed the ad verecundiam fallacy even if it were granted that the lecturer is a bona fide expert. The fallacy here is the dismissal of Herbert's argument, without replying to it, on the grounds that he is not an expert. For even so, he could have a good point that raises critical questions and merits a reply.

This case may serve to throw some light on the question of why the fallacy of inappropriate appeal to expert opinion would be called "argument from reverence or respect" (sometimes also translated as "argument from modesty"). Anyone who challenges the say-so of an expert can be attacked as being "immodest" or as showing insufficient respect for the authority of a genuine expert on the subject. This can be such a strong form of attack that it has the effect of virtually muzzling a participant in dialogue by suggesting that he has nothing to say about the issue that could be worth listening to at all.

Other abuses of appeal to expert opinion in argumentation include quoting an expert incorrectly, rendering the expert's opinion in a misleading way without even quoting her exact words, or even using phrases like "according to the experts," which do not name a specific expert source.[4] All of these kinds of abuses of appeal to expert opinion in argumentation could be called types of ad verecundiam fallacy.

The ad verecundiam fallacy seems to be not one single error but a number of different ways in which appeals to expert opinion in argumentation can go wrong. We can have one type of fallacious appeal where the person cited is not really an expert and another where the person is an expert but in the wrong field. Still another type of fallacy occurs where the expert is not named or otherwise specified exactly enough. Yet another type of failure occurs where the expert is named and is a genuine expert in a relevant field but her opinion is not what it is said to be.

Expert testimony, however, is regarded as a legitimate type of evidence to support an argument in the law courts. And generally, it seems that many appeals to expert opinion in everyday argumentation are of a nonfallacious sort. Many instances of such arguments are cited in Walton (1989a, chap. 7). It seems, then, that not all ad verecundiam arguments are fallacious. And if this is the case, we are left with the problem of determining when an argument of this type is fallacious and when not.

A further problem in identifying the ad verecundiam fallacy is whether it should be defined more narrowly as appeal to expert opinion or more broadly as appeal to authority. The former is a more narrow, cognitive way of characterizing the fallacy.

5. Complex Question

The fallacy of complex question (many questions) is the asking of a question containing presuppositions that the respondent is not committed to and that would look bad for him if he did concede them. The classical case is the following type of example.

Case 17

Have you stopped cheating on your income tax?

The respondent who does not want to concede having cheated on her income tax (presumably, the normal respondent, in most instances) immediately concedes such cheating once she answers yes or no to the question. Any direct answer—there are only two, in this case, yes or no—immediately incurs the respondent's commitment to the presupposition. Thus the question is a kind of trap. Instead of giving a direct answer, the respondent should question the question: "How could I stop, or continue, if I never did it in the first place?" It seems then that the committer of the fallacy is the asker of the question, who is trying to unfairly force a concession.[5]

An illustrative example of this type of question was given in the 1988 election campaign, when Ted Koppel asked the following question during an interview with Michael Dukakis (1988, 53).

Case 18

What is it about the Bush campaign that has absolutely nailed you to the wall?

The same problem is evident here. Any direct answer to the question by Mr. Dukakis would concede that the Bush campaign has absolutely nailed him to the wall—a conclusion of defeat.

In other cases, it seems that the use of loaded terms in the question is the problem. The following case is cited by Engel (1982, 124).

Case 19

What are your views on the token effort made by the government to deal with this monstrous oil crisis?

Here the respondent may or may not agree with the description of the oil crisis as "monstrous" or the description of the government effort as "token."[6] But any direct answer to the question would concede these descriptions as acceptable to the respondent.

These kinds of questions need not be fallacious in every context of dialogue. For example, if the question in case 17 were asked of a defendant in a trial who had just previously admitted cheating on his income tax, it would not be a fallacy. Such a question is only fallacious if the presuppositions in it have not been conceded already by the respondent and would be prejudicial to his side of an issue.[7]

What makes such cases instances of the fallacy of complex question is both the complexity of the question and also the way they are used in a dialogue to prevent the respondent from giving an answer without questioning the question itself.

6. Begging the Question

The fallacy of begging the question, also called petitio principii or arguing in a circle, occurs in an argument where a premise depends on the conclusion, or is even equivalent to it, in such a way that the requirement of evidential priority is violated. Evidential priority requires that the premises be better known or more firmly acceptable than the conclusion subject to doubt.[8] For example, in Euclidean geometry, the theorems are numbered, to indicate that a higher-numbered theorem can only be proved using premises that are lower-numbered (evidentially prior) theorems (Mackenzie 1980).

One common example cited by many textbooks was originally due to Whately (1836, 223).

Case 20

[T]o allow every man an unbounded freedom of speech must always be, on the whole, advantageous to the State; for it is highly conducive to the interests of the Community, that each individual should enjoy a liberty perfectly unlimited, of expressing his sentiments.

In this case, the premise and conclusion are expressed in different enough terms to perhaps obscure their identity, but really they both state pretty much the same proposition. Since they are the same, or equivalent propositions, one can't be evidentially prior to the other.

Another example, as noted by Hamblin (1970, 34) has been a staple of many textbooks.

Case 21

The context is a dialogue between a man, Smith, and his bank manager, where the manager asks Smith for a credit reference. Smith replies: "My friend Jones will vouch for me." The manager comes back: "How do we know *he* can be trusted?" Smith's reply, "Oh, I assure you he can."

In this case, one person is supposed to vouch for the reliability of the other. The reliability of the vouchee is in doubt, or being questioned, and some secure source, whose reliability is not in question, is needed to reassure this doubt. But if the reliability of the voucher is questioned, the reliability of the vouchee cannot be used to reassure this doubt, because it is itself in doubt, in the first place.

As far as purely formal considerations of deductive logic are involved, there is nothing wrong with circular arguments, even ones like '*A* therefore *A*.' This form of argument is deductively valid in the sense that if the premise is true, then the conclusion must be true, too. What is wrong with arguments that beg the question is that the conclusion is in question, or is subject to doubt, and premises used to resolve this doubt must be evidentially prior, that is, they cannot be themselves in question, at least to the same degree. Such a premise is useless to resolve the doubt, to secure a line of evidence that should be acceptable to the respondent as a basis for coming to rationally accept the conclusion.

Another type of case (Walton 1991a, 3), is also a favorite with the textbooks.

Case 22

God exists!
How do you know?
The Bible says so.
How do I know what the Bible says is true?
Because the Bible is the word of God!

This case is in the form of a dialogue, so it is a little easier to diagnose the fault of reasoning.

Presumably, the context of dialogue is that of an exchange between a proponent who is a believer and a respondent who questions or doubts the existence of God. Of course it is possible that this is not true of case 22, but normally it would be expected that the respondent is putting the whole religious point of view into question, including the Bible as a source of evidence that can be taken for granted as reliable. When the proponent cites as his premise 'The Bible says so,' the respondent, as we naturally expect on this interpretation, questions this assertion. But the fallacy comes in when the respondent answers this question in turn with his next assertion, 'The Bible is the word of God.' The problem is that this statement surely rests on the proposition that God exists, which was the very statement to be proven in the first place.

In this case then, the circle is a bit longer, but the fallacy of begging the question has again been committed because of the failure of the sequence of reasoning to meet the requirement of evidential priority. It is a failure or fallacy, presumably because of what we know or assume about the context of the dialogue between the two parties. The burden or task of the one party is to convince the other party rationally by appealing to evidence that will be adequate or sufficient to resolve the other party's doubts. To fulfill this burden, the first party must cite evidence that is, or could be, acceptable to the other party. To qualify as acceptable evidence, any statements cited as premises will have to meet certain general requirements, one of them being evidential priority. The fault of begging the question is therefore not a purely formal or deductive failure of reasoning but lies in the use of reasoning in a dialogue between two parties engaged in a purposive conversation.

If the context were somewhat different in this case, the circular reasoning would not necessarily be fallacious. For example, if the respondent were a committed Christian who unquestioningly accepts the Bible as the word of God but still has some wavering or marginal doubts about his faith, the argumentation in the dialogue could be successful in restoring his faith and removing his doubt. He might reply, for example, "Of course you are right. It does say so many times in the Bible. And the Bible is the revealed word of God. I accept that." In such a context, the argument could have been successful, and there would be no fallacy of begging the question committed.[9]

It appears, then, that the fallacy of begging the question is committed because there is a conclusion that is in question for one participant in a dialogue, and is supposed to be proved by the other participant, by citing premises that will prove the conclusion by removing the other's doubts. This job is not successfully accomplished by citing premises that are equally in doubt for the other party. That only "begs for" the proposition in question rather than doing the job of proving it, as required by the dialogue.

7. Hasty Generalization

One common interpretation of the fallacy of hasty generalization is that it is an inductive error (also called "insufficient statistics") where the size of the sample upon which a generalization is based is too small to support it properly. According to Campbell (1974, 148), the more varied the population in the generalization, the larger the sample should be. A small blood sample is usually sufficient, for example, because the composition of the blood does not normally vary much throughout a person's body. By contrast, the fallacy of insufficient statistics would occur in the following kind of case (Campbell 1974, 48).

> *Case 23*
>
> Eight men in a bar are polled to make a generalization about public opinion in an upcoming federal election.

This sample of respondents is simply too small (aside from its being biased, as well). For example, Salmon (1984, 58) treats hasty generalization, also called insufficient statistics or leaping to a conclusion, as an error of taking too small a sample on which to base an inductive generalization.

Other textbooks, however, have treated hasty generalization as the fallacy of applying a presumptive rule in an overly rigid or insensitive way that fails to take exceptions or qualifications into account. The traditional name for this fallacy is *secundum quid*, meaning "in a certain respect." For example, Joseph (1916, 589) defined the fallacy of secundum quid as the following error: "It consists in using a principle or proposition without regard to the circumstances which modify its applicability in the case or kind of case before us." Joseph gave the following example (589).

> *Case 24*
>
> Water boils at a temperature of 212° Fahrenheit; therefore boiling water will be hot enough to cook an egg hard in five minutes: but if we argue thus at an altitude of 5,000 feet, we shall be disappointed; for the height, through the difference in the pressure of the air, qualifies the truth of our general principle.

In this case, the fault is not the inductive failure to take too small a sample of instances on which to base the generalization. It is to overlook a specific qualification concerning the circumstances to which the general rule is meant to be applied normally as a rule of thumb for practical action, like cooking an egg.

One can easily see how the fallacy of secundum quid, or overlooking qualifications, is a common enough error in practical reasoning.

A standard type of case traditionally used to illustrate this type of fallacy is the following.

Case 25

Everyone has a right to his or her own property. However, Jones has been declared to be dangerous to the public when overcome by homicidal fantasies due to his schizophrenia. He is now asking you to give his rifle and shotgun collection back to him, even though he does not seem very coherent. Therefore you must give him the weapons.

The fallacy here is the failure to recognize that such a right is subject to qualifications and is defeasible in certain situations. To treat it too rigidly in drawing conclusions based on an absolutistic interpretation of the rule is to commit the fallacy of neglect of qualifications.

Both the inductive and presumptive fallacies, whether we call one or both of them hasty generalization, are clear enough as a common type of error of reasoning. But unfortunately, the terminological confusion in the textbook treatments does not end there. The same fallacies or similar ones are often treated under the heading of "accident" or "converse accident." For example, according to Copi and Cohen (1990, 100), the fallacy of accident is committed when we apply a generalization to special circumstances or cases too rigidly, where it does not properly apply. And when we commit the reverse error of wrongly applying a principle that is true of a particular case to "the great run of cases," we commit the fallacy of converse accident.

Although Aristotle's account of the secundum quid fallacy was quite clear and useful—for example, see *On Sophistical Refutations* 180a23–180b41) where he discusses general statements that need to have qualifications attached—what he wrote on accident has not been very helpful or clear as material for logic textbook writers (Walton 1990b). To a great extent, this is due to the doctrine-bound nature of the concept of accident in relation to Aristotle's theory of essential and accidental properties. This theory is not really suitable for explanation to introductory logic students in informal logic courses. The Port Royal account of the fallacy in Arnauld (1964, 259–60) is particularly confusion-generating in mixing different types of errors and in calling neglect of qualifications the *fallacia accidentis*.

The fallacy of hasty generalization, then, is in a particularly confusing and contradictory state, as presented in the various textbooks. One thing we need to do is to get away from the obscure and misleading terms "accident" and "converse accident" altogether. We also need to recognize that two important and distinct types of errors should be treated under the heading of the fallacy (or fallacies) of hasty generalization (Walton 1990a). One is the inductive error of generalizing inductively from too small a sample of evidence. The other is the presumptive error of neglecting qualifications to a presumptive rule in applying it to particular circumstances or exceptional cases. This second type of fallacy involves basically the same type of error whether the argument moves from the general rule to the specific case or vice versa.

What appears to be fallacious about this fallacy of ignoring qualifications is the failure to be flexible and open-minded enough in argumentation to recognize and allow for legitimate exceptions to a generalization when they arise (Walton 1992c, 282–84). It seems that many generalizations in everyday argumentation are of a nonstrict type that admits of exceptions.

8. Slippery Slope

The slippery slope fallacy occurs where one party warns a respondent that if he takes some contemplated course of action, it would trigger a whole series of ensuing events, unleashing an irresistible force that would result in some particularly horrible outcome for the respondent. An example from Johnson and Blair (1977, 166) concerned the proposal of the Canadian government in 1972 to issue work permits to Canadian workers. Dennis McDermott, the leader of the United Auto Workers, argued against the proposal.

> *Case 26*
>
> [The work permits] would run counter to our traditional freedoms and would be the *first step* toward a police state.

In this case, we can appreciate that the work permits would make it easier for the government to keep track of who is working where. But we are not told exactly how this would lead to a "police state." The idea of a police state sounds horrible, however, even menacing. It sounds so bad that if work permits would lead to it, then work permits sound like a bad idea.

But would work permits really lead to a police state? According to Johnson and Blair (1977, 166), the problem with McDermott's "brief causal story" is that we are given no idea what the intervening steps are. Moreover (166–67) it is easy to throw doubt on the argument by wondering why work permits would be any more of a threat to liberty than, say, drivers' licenses.

The slippery slope argument has often been illustrated by the famous domino argument used by Richard Nixon to warn against stopping the Vietnam war. Nixon argued that the fall of Vietnam would lead to the fall of other countries to Communist forces (Hardin 1985, 63).

Case 27

. . . would mean ultimately *the destruction of freedom of speech for all men for all time* not only in Asia but the United States as well. . . . We must never forget that if the war in Vietnam is lost . . . , the right of free speech will be extinguished throughout the world.

In this case, possibly the intervening steps could have been filled in. But in hindsight, the argument seems less than plausible. We can see that it was a weak speculative type of argument, even at the time, that was powerful more because it was scary than because of any strong evidence to back it up.

Another example of the slippery slope fallacy is a case more fully described in Walton (1992a, 195). In this case, two fundamentalist religious sects were disputing ownership of a territory, a holy mountain, held to be sacred by both of them. One side counseled moderation and sharing the mountain, but a fiery radical on the other side argued as follows.

Case 28

If we give this other sect even one centimeter, if we let them place even one toe on the Mountain, it will be the end of our holy places. We must ward off their attack on our holy place by dying a glorious death. Kill the infidels!

In this case, the slope is used as a tactic to rouse a polarized and quarrelsome viewpoint to subvert negotiations on the issue.

You can see a link between the slippery slope argument and the ad baculum argument in cases 26, 27, and 28. The slippery slope is used to exploit the fear or apprehension of the respondent that some horrible outcome suggested by the slope might come about. In slippery slope arguments, the argument is often questionable because some possible disaster may be sketched out roughly as an outcome of a proposed action, without any real proof's being given that this outcome will occur. The uncertainty of the future, however, combined with the disastrous or horrible outcome described, can be a powerful appeal to fear.

The slippery slope fallacy is also somewhat reminiscent of the secundum quid fallacy, because both may exploit or convey a rigid, dogmatic type of attitude that is not sufficiently flexible or sensitive to exceptions and qualifications in a given case. In case 28, the alternatives are presented in a rigid and polarized way that leaves no way open for discussions or qualifications. If the enemy places "even one toe" on the holy territory, then "it will be the end of our holy places."

The "enemy" is portrayed as relentless, inflexible, and not open to discussion or negotiation. This "us or them" attitude of dogmatic rigidity, or even fanaticism, seems to be part of what makes the slippery slope argument fallacious in this case.

As Walton (1992a) has shown, however, slippery slope arguments can be used correctly in some cases as a reasonable type of argument to shift a burden of proof in practical reasoning. Some of the correct kinds of slippery slope arguments are cited in chapter 5, section 10, below. Hence there is a problem of determining, in a given case, whether a slippery slope argument is fallacious or not.

9. False Cause

The fallacy of false cause (*post hoc ergo propter hoc*), according to Copi and Cohen (1990, 101), is "the error of concluding that an event is *caused* by another simply because it follows that other." Following is meant in the sense of "temporal succession" (101). A common example is given by Engel (1976, 93).

> *Case 29*
>
> Twenty-five years after graduation, alumni of Harvard college have an average income five times that of men of the same age who have no college education. If a person wants to be wealthy, he or she should enroll at Harvard.

In this case, the premise that there is a correlation between high income and graduation from Harvard may be quite true. Harvard, however, accepts only outstanding students, who tend to come from families of affluence and influence. As Engel notes (93) Harvard graduates would be likely to achieve high incomes no matter what college they went to or whether they went to college at all.[10] It does not follow that attending Harvard is the cause of the high incomes.

Another example is the following case, where Bob, a sixty-year-old, had just read a study published in the *Archives of Internal Medicine* (January 1992), involving eight hundred Michigan residents age sixty and over.

> *Case 30*
>
> **Bob:** A new study in Michigan found that older people (over sixty) who drink coffee are nearly twice as likely to be sexually active as those who don't. I'm going to start drinking a pot of coffee every day.

In this case, Bob's premise, citing the finding of the study he found in the medical journal, could be quite reasonable. The conclusion he draws from it, however, is not. According to the survey's principal researcher (as reported in *Newsweek*, January 29, 1990, 3), "no cause-effect relationship between coffee and sex has been proved. It may be that coffee simply stimulates the senses, or that sexually active people are generally uninhibited and like strong flavors." By leaping to the conclusion that he should take this kind of action, it seems that Bob is presuming a causal relationship where none may exist. This is the post hoc fallacy.

There is, in general nothing wrong or fallacious about arguing from a correlation to a causal conclusion. The error would seem to be one of leaping too quickly to such a conclusion without taking other factors into account that might defeat the inference (Walton 1989a, 212–34). In this respect, the fallacy could be seen as one of ignoring exceptions—perhaps a special case of the ignoratio elenchi fallacy or a case similar to it.

What is distinctive about the post hoc fallacy, however, is its causal nature. So far, neither science nor philosophy has been able to present any widely accepted analysis of the concept of a cause. It seems to be a practical idea, based on a "field," or on what is held constant (other things being equal) in a given case, where the production of one event results in another.[11] Whatever causality is, it is not established conclusively by a correlation alone.

10. Straw Man

According to Johnson and Blair (1977, 35), the straw man fallacy is committed when a participant in dispute misinterprets the opponent's position and then proceeds to argue against this (misrepresented) view: "When you misinterpret your opponent's position, attribute to that person a point of view with a set-up implausibility that you can easily demolish, then proceed to argue against the set-up version as though it were your opponent's, you commit *straw man.*" The tactic is to make your opponent's argument look bad by (wrongly) identifying it with a view that looks loathsome or dangerous to just about anybody.

For example, suppose that Mavis and Jim are arguing about improving the environment by controlling pollution, and Jim has argued for a moderate position on guidelines for industrial pollution. During the dialogue, Mavis argues as follows.

Case 31
The cost of making the environment a natural paradise on earth would be catastrophic for the economy of an industrialized country like ours.

The question in this case is whether any of Jim's commitments in the previous dialogue would justify describing his position as one of "making the environment a natural paradise on earth." If this is an exaggerated or distorted "set-up" version of Jim's position, Mavis has committed the straw man fallacy.

11. Argument from Consequences

The fallacy of *argumentum ad consequentiam* is the error of arguing that a proposition is false (true) on the grounds that the proposition (or the policy of carrying it out) would have bad (good) consequences. This type of argumentation is supposed to be a fallacy according to Rescher (1964, 82) because the premises "deal only with the consequences that are likely to ensue from accepting the conclusion, and not with its truth."

The following two examples from Rescher (82) illustrate the fallacy.

> *Case 32*
>
> Vegetarianism is an injurious and unhealthy practice. For if all people were vegetarians, the economy would be seriously affected, and many people would be thrown out of work.

> *Case 33*
>
> The United States had justice on its side in waging the Mexican war of 1848. To question this is unpatriotic, and would give comfort to our enemies by promoting the cause of defeatism.

In case 32, the problem appears to be partly one of relevance, of a shift in the issue. The conclusion, presumably, is that vegetarianism is injurious to one's personal health, whereas the premise cites bad social consequences that do not necessarily or directly relate to the personal health of the vegetarian. The other problem with this case is that it is simply a weak or unpersuasive argument. The premise that vegetarianism would throw many people out of work does not seem very convincing.

The problem with the second case is that the practical question of whether questioning one side would have bad consequences is really not relevant to the issue of which country in the war had justice on its side. Even if it were true that questioning the U.S. side would give comfort to enemies (which does not seem very plausible at this time), this point really does not bear on the question of which side was right or wrong in the war. As Rescher points out, both arguments seem to shift from truth to practical questions of consequences.

Still, the view that all arguments from consequences to the truth or falsity of a proposition are fallacious does not hold up. For example, suppose in case 32 it was argued that vegetarianism is an injurious and unhealthy practice because it has led people to get osteoporosis in many cases. This would be an argument from consequences, but it could be a reasonable argument. The difference appears to be that in the two fallacious cases there was a shift in the issue, so that the consequences cited were not really relevant. Getting osteoporosis, however, would be relevant to the issue of whether vegetarianism is unhealthy.

Another case will illustrate that when the fallacious type of argumentation from consequences occurs, it is because there has been a shift to a different type of dialogue.

Case 34

Two politicians are arguing about the issue of whether a woman should have the right to an abortion. The prolife politician argues against this proposition on the grounds that the fetus has a right to life. The prochoice politician replies: "If you take that view, you will not be elected."

In this case, the discussion shifted away from a critical discussion on whether abortion is right or not to a practical kind of advice-giving dialogue where one politician is warning the other about the political consequences of adopting a certain view. From the perspective of the first discussion, the shift almost looks like a kind of intimidation tactic, or a move to shut the other participant up. In light of this shift, the argument from consequences does seem to be a kind of fallacy, because it appears to have the effect of blocking the original critical discussion and diverting it to a different question.

On the other hand, suppose the consequence cited by the prochoice politician is in fact an accurate prediction and that the issue of "getting elected" is the more important goal in context. Then the shift from discussion of the abortion issue to a practical advice kind of dialogue, containing a warning, could be justified and appropriate.

Whether the argument from consequences is fallacious or not in this case, then, seems to depend on the context—and in particular on whether a shift from one type of dialogue to another is appropriate in the case. It seems that the evaluation of whether an instance of the argument from consequences is fallacious depends on contextual factors that determine whether a shift from one type of dialogue to another is reasonable or not.

Many slippery slope arguments—see especially case 27 above—are species of argumentation from consequences. It also appears that many ad baculum arguments—see case 6 above—are species of argumentation from consequences.

Argument from consequences is only featured in a small minority of current logic textbooks. Even so, it is an important fallacy in its own right because it is a very common kind of argument in everyday reasoning and because it is a type of argumentation that underlies some of the other major fallacies. The problem with it is to know when it is fallacious and when not.

12. Faulty Analogy

Arguments from analogy are frequently recognized by textbooks as being legitimate or reasonable kinds of arguments that can make a conclusion more or less probable or plausible. For example, Copi and Cohen (1990, 363) offer six criteria for evaluating analogical arguments as stronger or weaker in a given case. Many other textbooks cite fallacies in the use of analogical arguments, however, and argument from analogy, when it is not treated as if outright fallacious, is often treated as certainly a kind of argumentation that is tricky. It can be misleading. The misuse of analogical argumentation typically comes under a label like false analogy, faulty analogy, or misleading analogy.

According to Damer (1980, 49), the fallacy of faulty analogy "consists in assuming that because two things are alike in one or more respects, they are necessarily alike in some other respect." The following two examples are fairly typical illustrations of this type of fallacy (49).

> *Case 35*
>
> Smoking cigarettes is just like ingesting arsenic into your system. Both have been shown to be causally related to death. So if you wouldn't want to take a spoonful of arsenic, I would think that you wouldn't want to continue smoking.
>
> *Case 36*
>
> Suppose someone defended open textbook examinations with the following argument: "No one objects to the practice of a physician looking up a difficult case in medical books. Why, then, shouldn't students taking a difficult examination be permitted to use their textbooks?"

In the first case, the premise that both smoking and ingesting arsenic are related to death is true. Obviously there is an important difference, however: arsenic is much more toxic and is immediately fatal. Hence the comparison is an exaggeration.

The second case is an even poorer argument because there is very little similarity between the two cases apart from the act of looking inside a book for information, as Damer (49) notes. The purpose and context of the action in the two cases are quite different. The effort to use this argument to make a case for open book exams would be quite a weak attempt. It would be open to many strong objections that the two types of cases differ in important respects.

As these two cases illustrate, the problem with examples of faulty or false analogy suggested by the typical textbook illustrations is that the analogy is not altogether worthless, but it is simply too weak, or open to too many objections, to sustain much of a burden of proof in the argument. There is some question, then, whether these arguments are really fallacies in some strong sense or whether they are just weak or questionable arguments.

Of the two cases above, the open book example is perhaps the more questionable and the more subtly misleading comparison. Perhaps we could say that poor analogical arguments tend to be fallacious to the extent that they are more subtly misleading.

Another approach might be to say with Damer that an analogical argument is fallacious where it is assumed that the two things are necessarily alike or are alike in all respects. The fallacy here would be in pressing the analogy too hard or in trying to portray it as immune to critical questioning.

13. Linguistic Fallacies

Ambiguity is multiple meaning; for example, the word 'bank' could mean a savings bank or a riverbank, in different contexts. Vagueness is the lack of a clear cutoff point in the application of a word to a case. For example, some persons are definitely rich, and some are definitely not rich, but there exists a range of people that are borderline with respect to being rich. Of course a term like 'rich' can always be defined precisely—say, at assets of one million dollars—but then it could be argued that the definition is arbitrary, that it unfairly includes some and excludes others. Hence we often argue about terms and how to define them.

Ambiguity and vagueness can lead to problems in communication, but they are not inherently bad or fallacious in themselves. When they are conjoined to arguments in certain ways, however, fallacies occur of a kind Aristotle called "inside language." We could call these linguistic or verbal fallacies.

The fallacy of equivocation occurs where a word that is essential in an argument is used ambiguously in such a way that it makes the argument appear sound when it is really not. A sound argument is a valid argument with true premises.[12] The following example of an equivocal argument is given by Carney and Scheer (1964, 47), Kilgore (1968, 55), Frye and Levi (1969, 118), and Byerly (1973, 59).

Case 37

> The existence of a power above nature is implied in the phrase "law of nature," which is constantly used in science. For whenever there is a law, there is a lawgiver, and the lawgiver must be presumed capable of suspending the operation of the law.

In the first premise 'law' means a regularity or uniformity as described by an equation or statistical correlation in science. In the second sense, a law is a code of conduct, like a statutory or legislative law, set down by an authority, like a court. The first sense is the sense used in science. But in this sense, it is not true that there is some lawgiver behind the law who is capable of suspending it. This is true only in the other sense.

Once you disambiguate the word 'law,' it is easy to see that the argument is not valid. It only appears to be a valid argument with true premises when the ambiguity is masked. Once the ambiguity is revealed, the "one" argument is really several arguments, none of which is valid with true premises. Any attempt to make a univocal argument of it results in either a false premise or an invalid argument.

Amphiboly is the same kind of fallacy except that the fallacy is syntactical (structural or grammatical) rather than due to the ambiguity of a term. For example, the sentence 'Aristotle taught his students walking' is syntactically ambiguous. Two good examples of amphiboly are given by Michalos (1969, 366).

Case 38

> For sale: 1964 Ford with automatic transmission, radio, heater, power brakes, power steering, and windshield wipers in good condition.
>
> When you inspect the car, you find that the windshield wipers are the *only* accessories that are in good condition. When you charge the vendor with misrepresentation, he replies, "You misread the ad. Read it again."

Case 39

> The attendant at a roulette wheel in an amusement park offered some naive spectators "ten bets for a dollar." Since this sounded like a bargain, the spectators gave him the dollar. After the first bet was made and lost, they began to make a second. But the attendant insisted that they had misunderstood him. "Ten bets for a dollar," he explained "meant ten bets for a dollar *each*."

In both these cases, you would have to be quite naive to be taken in by the "spiel." But they indicate how amphiboly could be a serious fallacy in a more subtle case of business negotiation, say, where a legal contract is full of complex terminology and sentence structures. The same can be said for equivocation. It becomes a serious fallacy in a longer sequence of argumentation where a term gradually changes its meaning over several uses. In such cases, it might be difficult to spot the fallacy.

The same type of fallacy is called *accent* when the ambiguity arises from changes of emphasis given to the parts of a sentence. For example, Copi and Cohen (1990, 115) note that many different meanings can be given to the sentence 'We should not speak ill of our friends.' depending on which word is emphasized. Using small print and putting some words in large letters are physical ways of achieving the same effect and are often used in advertising.

Many textbooks—including Copi and Cohen (1990, 116)—also include *wrenching from context* as a type of accent fallacy. But this is a separate and serious verbal fallacy in its own right. In the fallacy of wrenching from context, words, phrases, or sentences are left out of a quotation, or disparate parts of the quotation are juxtaposed, so that the result invites a misleading interpretation, allowing the quoter (and his readers) to draw a conclusion that was not meant by the writer. This kind of fallacy can occur, for example, in using the say-so of an expert source to back up one's argument. The problem is one of interpreting a context of discourse fairly, to represent the writer's position accurately. This fallacy is related to the straw man fallacy.

Words and phrases in natural language always have positive and negative connotations, so it is always good in argumentation to look for key words that might be used by one side in a dispute and might be prejudicial to the view of the other side.

Case 40

Two historians are discussing which side was at fault in starting a war over some disputed territory. The one historian keeps describing the actions of the one side as "terrorist" and describes the other side as "freedom fighters."

Using these loaded terms or prejudicial words in such a way, in the context of the dispute, is an instance of the fallacy of question-begging epithet. The context of the dispute is the attempt to fix blame for starting the war. But calling the actions of the one side "terrorist" and the other side "freedom fighters" already predetermines the guilt of this side, closing the issue.

Of course, there is nothing fallacious or logically wrong per se with using language that has connotations that support your side of an issue. People routinely do so in everyday argumentation, and it is generally acceptable, at least up to a point, as part of partisan advocacy for one's point of view in a debate or critical discussion. But it becomes a problem if the language used is so strong that it prohibits the other party from putting forward her point of view, giving her no room to argue. Language so prejudicial that it is question-begging is open to challenge and should be questioned critically.

Language used in the wording of questions in public opinion polls can bias the results in a misleading way. In 1967, for example, two New York congressmen, Seymour Halpern, a Republican from Queens, and William Fitts Ryan, a Manhattan Democrat, polled their constituents on the conduct of the war.

> *Case 41*
>
> Halpern asked, "Do you approve of the recent decision to extend bombing raids in North Vietnam aimed at the strategic supply depots around Hanoi and Haiphong?" Sixty-five percent of those who responded said they supported the decision. At just about the same time, Ryan asked his constituents, "Do you believe the United States should bomb Hanoi/Haiphong?" By contrast, however, only 14 percent of his respondents supported bombing! The New York newspapers gave a great deal of attention to the two polls, in part because of the apparent conflict between them. Some people read the two surveys as confirming the theory that opposition to the war was coming essentially from upper-middle-class liberals, such as those in Ryan's district, while middle-class Americans, such as those living in Queens, were solidly behind the administration. [Wheeler 1976, 153]

The real difference in the responses can be traced to the differences in the wording of the questions. Halpern's question suggests only a defensive stance that is just an extension of already existing policies. Ryan's question, in contrast, appears to suggest an action that could be much more radical—bombing cities, a possible future action that could be quite a departure from existing policy. This question is worded so that it influences a cautious person, especially someone who may not know very much about the actual situation, to say no.

As Wheeler (154) comments, it is not clear whether Ryan and Halpern "deliberately loaded their questions or were simply oblivious to nuances in wording." But the potential for error, deception, and confusion is clear.

Most textbooks also classify the fallacies of composition and division as linguistic fallacies.[13] This practice seems hard to explain at first, because the examples typically given are arguments that have to do with the relationships between parts and wholes of physical and not with linguistic aggregates. We can see how such a tradition evolved historically, however, because Aristotle thought of the fallacies of composition and division as linguistic.

In the *De sophisticis elenchis* (166a 22), for example, Aristotle wrote that the expression 'A man can walk while sitting,' is true in a divided sense, meaning that a sitting man has the power to walk. But in the combined sense, the sentence is false, meaning that a man can walk-while-sitting. This is a syntactic ambiguity of sentence structure, and hence Aristotle was right to think of fallacies related to such ambiguities as "inside language."

In the modern view, however—so called in Woods and Walton (1989, 97)—the fallacies of composition and division have to do with parts and wholes, mainly of physical aggregates. Thus composition and division, construed according to the prevalent modern view, should no longer be classified as linguistic fallacies. They are more accurately and usefully seen as formalistic fallacies that have to do with inferences from parts to wholes and vice versa. The other major fallacy that is linguistic in nature is the *sorites* or linguistic subspecies of slippery slope argument. It is a type of argumentation that exploits the vagueness of a term in an argument, in order to refute that argument.

The sorites slippery slope argument is related to a species of argumentation sometimes called the argument of the beard in logic textbooks. This type of argument is a species of refutation of the form: a term used if your argument is too vague, and therefore your argument cannot be used to justify your conclusion. This type of argumentation, however, is not a fallacy. It is, in many instances, a reasonable kind of argument, even though it can be a badly used or weak argument in other cases.

Vagueness itself is not a fallacy. It is only when combined with arguments in certain ways that vagueness can be part of a fallacy or can lead to fallacies of various kinds.[14] Both vagueness and ambiguity can be problems in communication and can disrupt or prevent successful communication in some cases. Not every failure of communication is a fallacy, however. A fallacy is a failure of argumentation or a misuse of argumentation that blocks the goals of certain types of dialogue that are normative contexts of argumentation.

14. *Ignoratio Elenchi*

Ignoratio elenchi (ignorance of refutation) is the Aristotelian term for the fallacy of not proving what you are supposed to prove in a dialogue, namely the thesis (proposition) for which you have the burden of proof. Copi (1982, 110) calls this fallacy *irrelevant conclusion*, said to be committed "when an argument purporting to establish a particular conclusion is directed to proving a different conclusion."

Most textbooks call this fallacy irrelevance of some sort, like irrelevant premise; sometimes it is called the "red herring" argument. Copi (1982, 110) gives two examples.

Case 42

When a particular proposal for housing legislation is under consideration, legislators may rise to speak in favor of the bill and argue only that decent housing for all the people is desirable.

Copi (110) comments that "presumably everyone agrees that decent housing for all people is desirable." But that is irrelevant to the issue of whether this measure will provide it. Hence the argument is a fallacious ignoratio elenchi.

Case 43

In a law court, in attempting to prove that the accused is guilty of murder, the prosecution may argue at length that murder is a horrible crime and may even succeed in proving that conclusion.

Copi (110) comments that inferring the conclusion that the defendant is guilty of murder from these remarks about the horribleness of murder is to commit a fallacious ignoratio elenchi. In one sense, the premises are relevant to the conclusion in these arguments. The proposition that all people should have decent housing is related in subject matter to the conclusion that this measure will provide decent housing. Both propositions share the common subject matter of decent housing. Similarly in the other case, the topic of murder is shared by the premise and the conclusion. What does 'relevance' mean here then? This is a basic problem.

Copi (1982, 93) calls all thirteen of the eighteen informal fallacies treated in his text "fallacious of relevance" except for the five called fallacies of ambiguity (roughly what we call linguistic fallacies). This means relevance is a pretty broad category for Copi, covering, it seems, virtually any "failure to prove." But just about any fallacy could be called a "failure to prove" (in some sense).

But at the same time, ignoratio elenchi is treated as the unique fallacy of irrelevance or failure to prove. This treatment once again raises the question of what relevance is and makes one wonder whether it is a term that is being used in different senses.

The danger, as Hamblin (1970, 31) put it, is that the fallacy of ignoratio elenchi, or irrelevant conclusion, "can be stretched to cover virtually every kind of fallacy." As Hamblin noted (31), this fallacy has tended to become a "rag-bag," a wastebasket for any perceived failure or error of argumentation that cannot otherwise be explained or justified as objectionable or fallacious. We don't want to define relevance so broadly that virtually any perceived error of reasoning or objectionable argument can be classified as a fallacy because the premise is irrelevant (in some undefined sense) to the conclusion.

Copi and Cohen (1990, 106) clarify this point by noting that ignoratio elenchi is the fallacy in which an argument "misses the point without necessarily making one of those mistakes" identified with the other specific fallacies, ad hominem, false cause, and so forth (except for petitio principii). But this statement still doesn't answer the question of what irrelevance (or missing the point) is.

15. Conclusions

We can intuitively grasp, from the examples given in this chapter, that these fallacies are powerfully persuasive and deceptive types of argumentation that quite commonly trip people up in everyday reasoning and disputation. But pinpointing the precise error in each case appears to be a formidable job. For one thing, as we have repeatedly noticed, the so-called fallacy has similar counterparts that appear to be the same general type of argument but are nonfallacious. For another thing, the task of defining each of these types of argumentation in a clear enough way to classify them as distinctive species of arguments is a source of many difficulties.

The general problem is that informal logic lacks the precise guidelines provided by the structures of formal logic. Each of the types of argumentation identified with the various informal fallacies has a certain practical distinctness—we are familiar with how each of them is used in everyday argumentation as persuasive (and often deceptive) tactics. But we seem to be far from being able to evaluate such arguments as correct or fallacious, in a given instance, by appealing to some precise but general guidelines that we could systematically use to apply to the given data of that particular instance.

The problem can be highlighted by contrasting formal logic with the informal fallacies. In formal logic, we have clearly defined forms of argument. And these are applied to a given case to determine whether that instance of argument is valid or not. With the informal fallacies, we lack such general guidelines, provided by a general account of the structure of each type of argument.

Ultimately, in chapter 5, the argument forms, or argumentation schemes, as we call them, for identifying the types of argument associated with the informal fallacies will be given. This chapter, by itself, will not solve all our problems with the informal fallacies, but it is an important first step.

In the textbooks, we find formal fallacies as well as informal fallacies. One might expect that these so-called formal fallacies would be much easier or even trivial to analyze because formal logic already gives us an account of the appropriate form of argument for them. Curiously, however, as we will see, this does not turn out to be the case.

Now we have some intuitive grasp of what the most common of the informal fallacies featured by the logic textbooks are. But we have seen, in each case, that we lack any solid answer on how to identify, analyze, or evaluate the type of argument involved. We are a long way from being able decisively to evaluate given instances as fallacious or nonfallacious, using clear general guidelines based on a logical theory that can be applied to individual cases. We now turn to various requirements that have to be put in place as steps needed if we are to construct such a theory.

3

Formal Fallacies

According to Hamblin (1970, 195), the idea of distinguishing certain types of argument as "formal fallacies" is comparatively recent, dating from Whately (1836, bk. 3), who took up a hint from Aldrich.[1] But the earliest account of formal fallacies is that of one Cassiodorus (sixth century after Christ), who wrote a short chapter on paralogisms, or arguments that violate Aristotle's rules for the syllogism (Hamblin 1970, 194).

Nowadays many textbooks treat only informal fallacies, perhaps on the presumption that because formal logic has precise schemata or rules, there is no need to treat formal fallacies separately. Quite a few of the textbooks, however, do have a section on formal fallacies to balance off the usually more lengthy treatment of informal fallacies.[2] Certain types of deductively invalid forms of argument, like invalid syllogisms or invalid forms of inference in propositional calculus, tend to be featured as the leading "fallacies" in this category.[3]

Formal logic has logical forms or schemata, made up of constants and variables, and logical properties of these forms, like validity and consistency, can be determined by precise, mathematical methods, independently of what is substituted in for the variables in a given case. And of course the treatment of informal fallacies has always lacked just this formal precision.

One might expect, therefore, that formal fallacies are much more precise and clearly delimited and that the concept of fallacy is here very clearly defined. This is not so, however. The concept of a formal fallacy turns out to be about as elusive and difficult to get straight as its informal counterpart.

1. Affirming the Consequent

In many logic textbooks, we find a fallacy called *affirming the consequent*, which is said to be a formal fallacy. The following example is given by Salmon (1963, 27).

> *Case 44*
>
> Men, we will win this game unless we go soft in the second half. But I know we're going to win, so we won't go soft in the second half.

Salmon puts this argument in "standard form" as follows.

> If we do not go soft in the second half, then we will win this game.
> We will win this game.
> Therefore we will not go soft in the second half.

It is easy to see, however, that this form of argument is invalid by comparing it to a more obviously invalid one that has the same form.

> *Case 45*
>
> If Elvis Presley was assassinated, then Elvis Presley is dead.
> Elvis Presley is dead.
> Therefore Elvis Presley was assassinated.

This method of showing invalidity is the method Massey (1975, 64) calls refutation by counterexample—see section 2 below. Since the Elvis Presley argument is clearly invalid and yet has the same form as the "going soft" argument, we can say that the latter argument is incorrect (invalid) by virtue of its form. At least, that appears to be the presumption behind calling it a formal fallacy.

Stebbing (1939, 160) gives a similar example.

> *Case 46*
>
> Since he said that he would go to Paris if he won a prize in the sweepstakes, I infer that he did win a prize, for he has gone to Paris.

She restates the argument (160) in the following form (overlooking the point that the speaker only "said" the first premise and could have been lying).

> If he won a prize in the sweepstakes, he would go to Paris.
> He has gone to Paris.
> Therefore he has won a prize in the sweepstakes.

According to her, this argument is fallacious, because "he might have had a legacy, or been sent to Paris on business, or he might have grown tired of waiting to win a prize and gone to Paris whether he could afford it or not" (160). Also, we can see that the argument has the same form as the arguments in cases 44 and 45. The fallacy committed, she writes, is said to be known as "the fallacy of the consequent." This formal fallacy is said to be a fallacy because arguments have a certain form that it is deductively invalid.

The fallacy of affirming the consequent is a formal (deductive) fallacy that may be contrasted with the valid form of argument modus ponens.

Modus Ponens (valid)	*Affirming the Consequent* (invalid)
If A then B	If A then B
A	B
Therefore B	Therefore A

In a simple, concrete type of case—for example, let A be 'This egg drops.' and B be 'This egg breaks.'—there is no problem in this case of confusing one of these forms of argument above with the other. It is clear that the modus ponens instance is valid and the instance of affirming the consequent is not. But if you take a more abstract kind of case—say, where A is the proposition 'Virtue is a skill.' and B is 'Virtue can be taught.'—there is more real potential for confusion. It seems that the conditional used in the major premise could possibly or plausibly go either way around. In this type of case, then, an argument that has the form of affirming the consequent could seem valid, perhaps by virtue of its resemblance to modus ponens, which really is valid.

Much the same kind of observation could be made in connection with the parallel forms of argument, *modus tollens* and *denying the antecedent*. The fallacy arises because the invalid form of argument, when used in context, may seem valid because of its resemblance to (or confusion with) the counterpart valid form.

Modus Tollens (valid)	*Denying the Antecedent* (invalid)
If A then B	If A then B
Not B	Not A
Therefore not A	Therefore not B

Denying the antecedent is deductively invalid in the sense that it is possible for both premises to be true while the conclusion is false. But presumably because of its similarity to modus tollens, it could mistakenly be taken as valid in some instances, and hence it can be called a formal fallacy.

Here, then, we can see how the idea of a formal fallacy arose as a plausible category in the textbook treatments. It evidently is a type of fallacy that could be analyzed in purely formal terms, using formal structures of propositional logic both to show why it is invalid and why it appears to be valid.

Seeds of doubt can be sown, however, with respect to this tidy and apparently attractive doctrine of formal fallacies. One is the well-known asymmetry thesis of Massey (1975), which warns us that just because an argument has an invalid form, it is not necessarily an invalid argument. Any argument has many forms, and even a valid argument has invalid forms. For example, an argument having the valid form of modus ponens also has the invalid form '*A, B;* therefore *C.*' Much depends, it appears, on how explicitly that form represents the structure of an argument, where the argument is in natural language.

For example, Capaldi (1971) portrays denying the antecedent as a confusion between a correct and an incorrect argument.

Case 47

If you take cyanide, then you will die. You take cyanide. Therefore you will die.	If you take cyanide, then you will die. You do not take cyanide. Therefore you will not die.

Capaldi (167) diagnoses the argument on the right as an "example of poor or fallacious reasoning," and the one on the left as a "true example of causal reasoning." Capaldi reasons that there are other causes of death besides taking cyanide, therefore the fallacy of denying the antecedent is the fallacy of believing that something (in this case, cyanide) is a necessary condition of something else (in this case, death), when "in actuality" it is a sufficient condition.

This analysis of this so-called formal fallacy as being (at least in part) a causal fallacy sows seeds of doubt about the purely formal nature of the fallacy. Capaldi's analysis is very similar to an analysis given by Aristotle of a fallacy he called the refutation of the consequent in *De sophisticis elenchis* (1955, 167b2–167b12).

> The refutation connected with the consequent is due to the idea that consequence is convertible. For whenever, if *A* is, *B* necessarily is, men also fancy that, if *B* is, *A* necessarily is. It is from this source that deceptions connected with opinion based on sense-perception arise. For men often take gall for honey because a yellow colour accompanies honey; and since it happens that the earth becomes drenched when it has rained, if it is

drenched, we think that it has rained, though this is not necessarily true. In rhetorical arguments proofs from signs are founded on consequences; for, when men wish to prove that a man is an adulterer, they seize upon the consequence of that character, namely, that the man dresses himself elaborately or is seen wandering abroad at night—facts that are true of many people, while the accusation is not true.

According to Aristotle's account, this fallacy arises from a switching around of the conditional. Take his example of the conditional "if x is honey, then x is yellow." The fallacy arises where the conditional is used in the first premise in the following kind of argument.

Case 48

If x is honey, then x is yellow.
This substance x is yellow.
Therefore this substance x is honey.

This fallacy could arise in a case, for example, where the substance x is really gall but only seems to be honey because it is yellow. You can easily see, however, that although the fallacy in this case can be portrayed as a formal fallacy of affirming the consequent, that purely formal analysis does not get to the heart of the problem. The heart of the problem is the mix-up between necessary and sufficient conditions, the reversal of 'If A then B.' and 'If B then A.'

What really creates (and explains) the fallacy in Aristotle's example is that something's being yellow might, in certain typical situations of inquiry, be a sign of its being honey but a fallible sign (subject to exceptions). And something's being honey could be an even more reliable indication that it would normally be yellow. Even here too, though, the conditional is subject to qualification; for example, if it is buckwheat honey, it might be brown and not yellow. So the argument from sign is a defeasible argument, expressed by a conditional that is meant to be subject to qualifications. But it is much weaker when it goes more one way than when it is taken to go the other.

So analyzed, then, the fallacy of consequent is not a purely formal fallacy. It has significant informal elements of how you interpret the conditional in natural language argumentation.

2. Invalidity and Fallacy

Most logic textbooks give a set of rules for valid syllogisms, and many of them identify certain formal fallacies with a breach of one of these rules. The six rules given in Whately's *Elements of Logic*, for example, are reprinted in Hamblin (1970, 196–97). The first rule is "Every syllogism has three, and only three, terms," and the so-called *fallacy of four terms* refers to the following type of case, from Whately (1836, bk. 2, chap. 2, pt. 2).

Case 49

Light is contrary to darkness.
Feathers are light.
Therefore feathers are contrary to darkness.

But as Hamblin notes (197), this example is not one of four terms, but equivocation. Or to put it another way, there really is no difference between the fallacy of four terms and the fallacy of equivocation. For case 49 is a clear, even a paradigm, case of the fallacy of equivocation.

The problem in case 49 is that because of the ambiguity of the term 'light,' and its use in one sense in one premise and in a different sense in the conclusion, we really have four arguments here. It only appears that we have one syllogism on the surface. Really, there are four arguments, and none of the four has true (plausible) premises and is valid. This is a typical case of the fallacy of equivocation.

This case is symptomatic of what Hamblin goes on to demonstrate generally (191–205). No matter how you try to patch up a system of classification so that each formal fallacy corresponds to a violation of one of the rules for a valid syllogism, the project fails. Hamblin (203) also thinks that comparable attempts to give a small set of rules used to proscribe the fallacies of 'affirming the consequent' and 'denying the antecedent' are "too fragmentary" and "strangely ill-judged." The whole project of trying to define or classify formal fallacies as violations of some sets of rules that define validity for systems of formal logic like syllogistic or propositional logic seems not to work.

Another consideration will illustrate how such a project tends to fail despite its initial plausibility and attractiveness to textbook writers. Hamblin (200) gives an example of an invalid syllogism that breaks all the rules of a particular set of three rules for a valid syllogism even though violations of each of these three rules individually defines a particular formal fallacy. Hamblin draws the following conclusion (201).

What this means is that, although the set of three rules is quite adequate to define validity and hence formal fallacy, it does not give us a classification of fallacies, in the sense of a division into mutually exclusive categories; unless we are content to count each possible combination of ways the rules may be broken as generating a different category, in which case there would be seven categories altogether.

Hamblin goes on to demonstrate the extreme difficulty of getting a set of rules that are individually necessary and jointly sufficient for validity of syllogisms that can, at the same time, serve as a system of mutually exclusive categories for classifying formal syllogistic fallacies. His conclusion is that such a system is not useful because it is ad hoc and arbitrary and that perhaps we should "give up trying to produce a classification of formal fallacies altogether" (203). Moreover, he thinks that this conclusion is reinforced by considering what would be involved by producing a similar system of formal fallacies for modern logic, that is, propositional logic and quantifier logic.

These remarks are very disheartening for the doctrine that formal fallacies can easily or straightforwardly be defined as violations of sets of rules defining formal validity for a type of argument. There seems to be a widespread assumption generally that formal fallacies are in much better shape than informal fallacies because, at least in formal logic, we have clear sets of rules that define exactly which arguments are valid and which are not. And therefore, goes the presumption, there is really no (comparable) problem in defining, analyzing, or classifying formal fallacies. But as attractive as this presumption seems, it seems to collapse when any serious attempt is made to carry it out. It seems that the fallacy is not just the possession of an invalid logical form by a given argument. Something else seems to be involved as well in the concept of a fallacy, formal or informal.

These cautionary remarks about the trickiness of the concept of formal fallacy are reinforced by the asymmetry thesis of Massey (1975), according to which the use of formal logic to prove validity is very different from its use to prove invalidity of an argument. According to Massey, the logic textbooks teach us how to use a system of formal logic like propositional calculus or quantifier logic to prove the validity of a given argument in natural language. We paraphrase the argument (Massey 1975, 63) by transforming it into an argument form of a logical system the theory of which we recognize as correct, and then we test the resulting argument form for validity. If the form is valid, then the original argument (said to have that form) is declared valid. This method works essentially because of the principle of uniform substitution of constants for variables, guaranteeing that if a form is valid, *every argument having that form is valid.*

The same technique for proving invalidity would not work, however. Basically, the reason is that any given argument has many different forms, and an argument that has a valid form also has invalid forms. So an argument could have an invalid form (in some system), but that does not mean it is (necessarily) invalid. It might also have a valid form (in the same or in a different system) as well.

Massey concludes from these observations that there is no way that formal logic can be used to prove that an argument is invalid. Massey (64) agrees that there is one wholly nonproblematic way to show that an argument is invalid. That is to show that the premises are all true and the conclusion false, in a given case. But Massey objects (64) that this method can rarely be applied, because in fact we may not know how to prove that the premises are true and the conclusion is false. Moreover, even if we did know, this procedure takes us beyond formal logic to material questions of truth and falsity, which in most cases, are not questions for logic to decide. Massey's asymmetry thesis has implications for the presumption that you can use some set of rules that define formal validity of a class of arguments to classify or define formal fallacies relative to that type of argumentation.

We shouldn't draw the conclusion that formal logic is useless in the identification or evaluation of formal fallacies or that formal fallacies do not exist as a class. But we need to have a clearer idea of how formal logic is correctly used for such a purpose. What is crucial in such an application of logic is that the given argument to be evaluated occurs in a natural language setting. It has to be translated or paraphrased into a logical form made up of constants and variables in an artificial language. But there are serious questions, in any given case, about how the form represents the structure of the given argument. It is these questions that will be important in determining whether a fallacy has been committed or not in the given case. In a sense, what is shown is that formal fallacies are case-oriented—relative to the particulars of a given case and the context of dialogue. It is not just informal fallacies that have this case-relative aspect.

The traditional slogan is that a fallacy is not only an invalid argument but one that seems valid. Perhaps this 'seems' is not a psychological property but a question of how the argument was used in a context of dialogue as judged by the wording of a given case. What follows from this is that supposedly formal fallacies like the fallacy of consequent may have a formal aspect but may turn out to be less than purely formal in certain key respects, once we see how logic is used rightly to identify and analyze them.

3. Consequent as Fallacy

The so-called fallacy of consequent, or affirming the consequent, has turned out to be a lot less clear as a formal fallacy than we might initially have thought. For it seems to be only part of what makes this species of argumentation fallacious in a given case to show that it is an instance of the formally invalid argument scheme 'If A then B; B;

therefore *A*.' Perhaps a discussion of one of the other cases cited by Aristotle will help to make this point clearer. In the quotation above, Aristotle cited the following simple case.

Case 50

> The earth becomes drenched when it has rained.
> The earth is drenched.
> Therefore it has rained.

Part of the problem with this case that could lead us to classify it as an example of the fallacy of consequent is that, in context, it is natural for us to see the drenching of the earth as an event to be explained. And it is also natural, again in context, to cite rain as the cause or explanation of the drenched earth. So what leaps out at us when we encounter case 50 as an argument is to see it as an instance or use of argumentation from sign, as below.

Case 50a

> I see the drenched earth.
> What I see (the drenched earth) is a plausible sign that just previously (although perhaps I did not see that), it just rained, assuming I know of no better explanation.
> Therefore it is a reasonable presumption that it has rained.

The conclusion can be taken as a defeasible presumption subject to correction if some other cause of the wetness of the earth comes to be known in this case. According to this natural interpretation, the argument can be taken as a nonfallacious argument from sign.

But what makes the argument cited by Aristotle fallacious? Presumably, the clue is to be sought in Aristotle's remark, "if it is drenched, we think it has rained, though this is not necessarily true." The fallacy is taking the major premise as a strict conditional of the form, 'In every case (without exception) if the earth is drenched, then it has rained.' For given the premise, 'The earth is drenched.' we can derive the conclusion 'It has rained.' by modus ponens.

The fallacy, then, is that the argument is taken as a strict, deductively valid argument, subject to no exceptions, whereas it should only be taken as a presumptively reasonable argument with the major premise, 'If the earth is drenched, then presumably (in this case) it has rained.' Really then, the fallacy is one of hasty generalization or secundum quid (neglect of qualifications).

In fact, what has been shown is that the fallacy of secundum quid is, to some extent at least, a formal fallacy, concerned with strict conditionals. And at the same time, if this case is an instance of consequent, then consequent is, at least partly, an informal fallacy.

Furthermore, Aristotle claims, the fallacy arises because of "the idea that consequent is convertible." What he seems to mean here is that the conditional 'If it rains, then the earth is drenched.' may seem stronger, or less subject to defeat, than the conditional reversed, which was the major premise above. So here we seem to have two explanations of the fallacy. One is the reversing of the conditional. The other is the confusion between a strict and a presumptive conditional.

Another very good clue to what this fallacy is can be gotten from Aristotle's third example of the presumed adulterer. If a man dresses very well and is seen wandering around at night, such signs could raise the suspicion that he is an adulterer. That reasoning, in itself, is not necessarily fallacious. Someone who wants to press forward an accusation of adultery, however, could "seize on" (to use Aristotle's expression) these facts and press strongly for the conclusion that this man is an adulterer. Where is the fallacy in this? Once again the clue is in Aristotle's suggestion that such facts are true of many people who are not adulterers. The argument could be fallacious where pressed ahead too hard so as to get a respondent to accept the accusation without giving the accused any benefit of doubt. It seems to be a classical case of secundum quid.[4]

The adulterer case is actually a classic case of an informal fallacy, as we defined this concept in chapter 2, because it is clearly a case where a presumption has gone forward in a dialogue with a failure of burden of proof appropriate for the conclusion drawn. The fallacy is that the one party has "seized on" the signs of possible adultery and pressed ahead too uncritically to draw or force a conclusion that is not adequately supported by two such defeasible signs. In the dialogue the benefit of the doubt should be given to the presumed adulterer and not the other way around, to his accuser. This is the heart of the problem, so formal logic is involved, but really the key to it is the informal fallacy of secundum quid involved in the shifting of a presumption in dialogue.

There are two conclusions to be drawn. One is that formal fallacies and informal fallacies are much more mixed in practice than the traditional treatment ever suggested. The other conclusion is that determination of a supposedly formal fallacy like consequent in fact involves essential linguistic and contextual elements of the argumentation used in a given case. Calling it a formal fallacy makes it seem more abstract and less contextual, but this appearance quickly proves to be an illusion once you try to show that such a fallacy has actually been committed in a real case.

None of this is to deny the usefulness of formal logic in analyzing some cases of fallacies. It is to say that even the so-called formal fallacies are less purely formalistic in nature than tradition has assumed. But some of them are well worth knowing about and are well worth teaching in textbooks.

4. Scope Confusion Fallacies

One important type of formal fallacy, called by Mackie (1967, 172) the *fallacy of rearranging operators,* has to do with reversing operators or otherwise confusing the scope of an operator in a proposition. Mackie (172) cites the following invalid inference as an illustration.

Case 51

It is certain that someone will win.
Therefore there is someone who is certain to win.

The premise could be symbolized as '$(C)(\exists x) Wx$' and the premise as '$(\exists x)(C) Wx$,' according to Mackie, but the inference is "facilitated by the fact that 'Someone is sure to win.' is ambiguous between the two." (172). In such a case, the fallacy could be seen as a formal fallacy, but it could also be seen as an instance of the informal fallacy of amphiboly.

This idea is somewhat disturbing, in relation to the idea of classifying between formal and informal fallacies. Yet it is clear that there is some sort of significant fallacy identified here, whatever you call it or however you classify it.

Indeed DeMorgan (1847, 247) identified this very same type of fallacious inference but classified it as an ambiguity of construction "in our language" under the heading of fallacia amphiboliae, citing the following case (247).

Case 52

It cannot, for instance, be said whether 'I intend to do it and to go there tomorrow' means that it will be done tomorrow or not. It may be either—(I intend to do it and to go there) tomorrow, or—I intend to do it and (to go there tomorrow). The presumption may be for the first construction: but it is only a presumption, not a rule of the language.

DeMorgan's example, like Mackie's, is quite a good one in that it represents a kind of confusion of reasoning that seems quite likely to be both common and also a significant type of error in everyday argumentation. Both cases have to do with confusion relating to the scope of an operator in a sentence.

In DeMorgan's case, it is the placing of the parentheses to indicate the scope of the operator—whether it applies only to the first proposition or to both—that is the problem. In Mackie's case, it was a reversing of the operators that was the problem. But both are questions of scope of operators in a sentence.

In the DeMorgan case, you could say that the problem is one of ambiguous punctuation. In fact, DeMorgan uses as another illustration of the same fallacy—the confusion between (3 + 4) × 10 and 3 + (4 × 10). Small wonder that he classified such fallacies generally under the heading of amphiboly.

Fallacies that are the result of careless construal of the scope of modal operators have been cited as the basis of specious arguments for deterministic conclusions. Mackie (1967, 172) cites the following inference.

Case 53

> Necessarily you will either go or you will stay.
> Therefore you will go necessarily or you will stay necessarily.

Here the fallacy is one of bringing the necessity operator (*L*) into the disjunction, of inferring from $L(A \lor B)$ to $LA \lor LB$ illicitly.

According to Thomas (1970, 143) a similar fallacy resides in the following form of argument for determinism.

Case 54

> Given the factors which caused a man to act as he did existed, he could not have acted otherwise.
> Those factors existed.
> Therefore he could not have acted otherwise.

Thomas (146) analyzes the fallacy as a verbal confusion between two inferences.

If A then LB	L (If A then B)
A	A
Therefore LB	Therefore LB

The inference on the left is valid, but the one on the right is invalid. The problem is that the first premise of the argument in case 54 is ambiguous—it could be taken either way.[5] The reading on the right, however, is the more plausible interpretation, perhaps. Thus the argument in case 54 is a fallacy—it looks valid but really it is not (when the major premise is construed in the more plausible and defensible way). Clearly this is an important type of fallacious reasoning.

The problem is what to call these cases. Are they instances of the formal fallacy of mismanaging the scope of an operator, or are they instances of the informal fallacy of amphiboly? You could call them either. Amphiboly emphasizes the linguistic element of ambiguity.[6] The formal classification shows that you can actually disambiguate the sentence construction using punctuation and operators in a formal system of logic.

The classification proposed here is that we call them cases of amphiboly while recognizing that in these two cases we are fortunate enough to be able to express the ambiguity in a formal, well-systematized, logical notation. This means that sometimes amphiboly is at least partly a formal kind of fallacy.

This conclusion will be rejected by those who insist on a sharp distinction, with no overlap, between formal fallacies and informal fallacies. This is an interest of vocal and influential groups, it must be remembered. Formal logicians tend to want to keep formal logic pure and free of contextual-pragmatic elements. Those who favor informal logic very often reject formal logic as a useful tool for the analysis of argumentation and fallacies. Hence there are strong motivations to keep formal fallacies and informal fallacies apart, on both sides.

Inconsistency is another bone of contention. It has long been defined as a purely formal concept by logic textbooks, but some, like Rescher (1987), have described it as a fallacy.[7] Inconsistency is clearly related to fallacies, so some clarification of it in this connection is needed.

5. Inconsistency

Inconsistency is not a fallacy per se. But inconsistency in an arguer's collective set of commitments in a dialogue is subject to critical questioning. For a set of commitments that are collectively inconsistent could not all be true. Thus in general it is a requirement of the tenability of one's position in argumentation that it be represented by a set of commitment propositions that are internally consistent. Therefore, it is a common, powerful, and inherently reasonable form of refutation in dialogue for one party to attack an opponent's argumentation on the grounds that the opponent's position (set of commitment propositions) is inconsistent.

Mackie (1967, 176) has expressed this point succinctly.

> A position or a system of thought cannot be sound if it contains incompatible statements or beliefs, and it is one of the commonest objections to what an opponent says that he is trying to have it both ways. Inconsistency has many possible sources, but one that is of special importance in philosophy is the case in which a thinker, in order to solve one problem or deal with a particular difficulty, denies or qualifies a principle he has previously adopted, although in other contexts he adheres to the principle and uses it without qualification.

As Mackie notes (176), inconsistency may not usually be obvious but may be concealed in a context of dialogue. A speaker may express a general principle in one context but then later, when the topic has shifted, deny it without realizing it. Inconsistency may also be concealed (Mackie, 1967, 176) "by the use of different expressions with a single meaning." In this type of case, the error could be partly a linguistic fallacy and not a purely formal error.

The concept of an explicit inconsistency can be defined in a purely formal manner: it has the form of a proposition A, conjoined to its negation, *not-A*. Some inconsistent sets of propositions, ones we might call *logical inconsistencies*, can be reduced to this form by logical deductions, for example, in propositional logic.

Other inconsistencies (in a broader sense of 'inconsistent') involve presumptions about the meaning of terms in natural language. For example, 'Jan is a bachelor.' and 'Jan is not male.' are inconsistent, given the presumption that 'bachelor' means 'male, unmarried person.' This type of inconsistency is different from logical inconsistency, as more narrowly construed above. The set of three propositions 'Jan is a bachelor.,' 'Jan is not male.,' and 'All bachelors are male.' is logically inconsistent, however. The third and first propositions together logically entail the negation of the second.

An argument with an inconsistent set of premises is formally valid, at least in classical deductive logic. But this is only a reflection of the way deductive validity is defined. A valid argument is one where it is logically impossible for the premises to be true and the conclusion false. Hence any argument where it is impossible for the premises to be true is, ipso facto, a valid argument.

An argument of this kind, however, will be of no use to convince any rational respondent in a dialogue that the conclusion is true, assuming he realizes that the premises are inconsistent. Such an argument does not give any rational support to its conclusion.

Attacking someone's argument on the grounds that its premises are inconsistent is a common form of refutation or dissociative argumentation. Quite often, however, the inconsistency is not an explicit logical inconsistency but depends on presumptions to which the arguer criticized is assumed to be committed, even though he never said so explicitly in so many words.

Quite often, such an inconsistency is a *pragmatic inconsistency*, meaning that it can be reduced to a logical inconsistency only by drawing implicatures, as opposed to strict logical implications, based on Gricean maxims of collaborative politeness in a dialogue.[8] Quite often, for example, a person's actions, as conceded by him in a dialogue, may be pragmatically inconsistent with certain goals or general policies he has also advocated in the same dialogue. This shifts a weight of presumption against his side to resolve or explain the inconsistency, if challenged.

In such cases, it is important to realize that the inconsistency alleged is not a logical inconsistency, in the sense defined above. To reduce it to a logical inconsistency, assumptions have to be made concerning the pragmatic implicatures of speech acts in a dialogue. Thus when this form of attack is judged to be fallacious, it is not a formal fallacy that is involved (at least not exclusively).

A formal element, however, lies at the basis of this kind of refutation. As Mackie (1967, 176) put it, "It is a formal fallacy to suppose that because your opponent has tried to have it both ways, he cannot have it either way—that every part of an inconsistent position must be false." This type of argumentation is indeed a fallacy, and it has formal elements, but is it a formal fallacy?

The principle behind this type of argumentation is that the falsity of a proposition A does not follow from someone's having maintained both A and its negation, $\sim A$. In fact the form of inference '$A \cdot \sim A$; therefore $\sim A$' is valid in classical deduction logic, however. Thus while formal elements are involved, it may be too strong to call the kind of argumentation cited by Mackie a formal fallacy.

This type of argumentation does seem to be a recognizable fallacy of some sort, however. It would appear to be very similar to (if not identical to) the analysis of the ad hominem fallacy given by Barth and Martens (1977). According to their analysis, the ad hominem fallacy is committed when it is argued that a proposition is true (absolutely defensible) if it follows from one's opponent's concessions in a dialogue.[9] The negative case of the fallacy would be to argue that a proposition is false (absolutely refutable) if its negation follows from one's opponent's concessions in a dialogue. The fallacy, so conceived, is the confusion between absolute (strong) refutation, meaning a showing that a proposition is false, and relative (weak) refutation, meaning a showing that a proposition is inconsistent with an opponent's concessions. In the second case, the refutation is relative to that opponent or to his commitments in prior dialogue.

This is a broad way to define ad hominem argumentation. But it does seem to have an important precedent. As Hamblin showed (1970, 160), the most likely origin of the argumentum ad hominem, under that name, as a distinctive type of argumentation (possible references in Aristotle excepted) is in Locke's *Essay* (1961). Locke defined the argumentum ad hominem as a way of prevailing on an opponent's assent, by pressing "a man with consequences drawn from his own principles or concessions."[10] If we take Locke to mean "logically deducible consequences," we get the type of ad hominem defined above and modeled in Barth and Martens's analysis. So conceived, the fallacy may not be a purely formal fallacy (perhaps), because it involves the key dialectical concepts of an exchange between two participants in an argumentative dialogue and of an opponent's set of concessions in that dialogue. But it certainly involves formal elements quite heavily, being based on the idea of an inconsistent set of propositions, as compared to a false proposition.

Inconsistency, then, is a formal concept (or at least the core concept of logical inconsistency can be defined in a formalistic way) that is not itself a fallacy but is nevertheless closely connected to important fallacies. And inconsistency generally is a very important idea for argumentation. Generally, inconsistency is a subject for criticism in argumentation, if found in an arguer's set of concessions or in premises he advocates as a basis for drawing a conclusion to be accepted, as shown true by those premises.

Curiously, Mackie (1967) classified inconsistency as a fallacy but as a fallacy in discourse, not as a formal fallacy. Most superficial observers would probably be inclined to presume that inconsistency is a formal fallacy if it is any kind of fallacy at all. But one can easily appreciate why Mackie classified it as a fallacy in discourse (an informal fallacy, in effect). For the inconsistency in question has to be one in an arguer's commitment set for the term 'fallacy' to be appropriate to apply to it.

Even so it seems better to say that what is fallacious is not the inconsistency per se but how it is managed, dealt with, or deployed in a dialogue exchange between two arguers. Refusing to acknowledge or deal with an inconsistency in one's position, once it has been pointed out by a partner in argumentation, could be a kind of rigidity or uncooperativeness that might be called some sort of fallacy. Or the Barth and Martens type of move, declaring an opponent's thesis absolutely false because it conflicts with some other proposition in his concessions, can also rightly be described as a fallacy.

On balance, then, it seems better to go against Mackie, and not classify inconsistency, at least by itself, as a fallacy. And if the real fallacy is how the inconsistency is mistreated in a dialogue, then there would seem to be a good case for classifying fallacies arising through the misuse of inconsistencies in argumentation as informal fallacies, as opposed to (purely) formal ones. All of this depends, however, on how you define 'formal.'

6. Composition and Division

The modern interpretation of the fallacies of composition and division takes them to be invalid arguments from parts to wholes and conversely, and typically physical parts and wholes are meant. Composition is arguing from properties of parts to properties of wholes, and division is the converse.

The classic illustration is the following (Copi 1982, 125), said to be a "particularly flagrant example" of the fallacy of composition.

Case 55

Every part of a certain machine is light in weight.
Therefore the machine, as a whole, is light in weight.

Another example (Copi 1982, 125) of the fallacy of composition has a somewhat different structure.

Case 56

A bus uses more gas than a car.
Therefore all buses use more gas than all cars.

The fallacy in this case has to do with an ambiguity in the word 'all.' Each bus uses more gas than each car, so you could say that in a distributive sense of 'all,' it is true that all buses use more gas than all cars. There are many more cars than buses, however, so that collectively speaking, all the cars use more gas than all the buses. Hence the fallacy in case 56 seems to combine composition with a kind of equivocation on two senses of the word 'all.'

In the premise, the phrases 'a bus' and 'a car' means that each bus has a certain property, as does each car. In standard quantification theory, we would render this as follows: 'For all x, if x is a bus, then for all y, if y is a car, x uses more gas than y.' In other words, if you take each bus and car pairwise, the bus will, in every case use more gas than a car. The word 'all' in the conclusion, however, refers to the collection or aggregate of buses, en masse, as it were. This whole aggregate of all the buses together, it is said, uses more gas than the aggregate of all the cars.

Neither of these statements is free from ambiguity, however. You could, perhaps even more charitably, interpret the premise as saying that generally, but with exceptions, a bus uses more gas than a car. Even so, the argument would be fallacious, assuming that the conclusion represents an aggregate or collective use of 'all.' You could also say that the conclusion is ambiguous, though. Perhaps it could be interpreted in the distributive (quantifier) sense of saying that every bus uses more gas than any car.

So we could say that this case is partly a formal fallacy, turning on the quantifier-word 'all,' partly a composition fallacy involving an aggregate sense of the phrase 'all buses,' and partly a linguistic fallacy of amphiboly.

More usually, cases like 55 are cited in the textbooks under the heading of the fallacy of composition. Cases like 56 tend only to be cited in a few textbooks where composition is given a fuller treatment.

Division is just the opposite fallacy. For example, the argument in case 55 could be turned around, putting 'heavy' in for 'light' in the premise and conclusion. Or another typical example is the following, from Copi (1982, 126).

Case 57

A certain corporation is very important.
Therefore Mr. Doe, a member of this corporation, must be very important.

It is easy to see how these standard examples of composition and division are fallacious arguments. But in comparable cases, it is less easy to dismiss this same type of argument as fallacious.

Case 58

All the parts of this machine are made of iron.
Therefore this machine is made of iron.

It seems that some properties are composable and divisible while others are not. The least we can say, at any rate, is that it should not be taken for granted that any property composes from the parts to the whole, or divides from the whole to the parts.

At first it appears to be something of a puzzle that composition and division are typically treated in the logic textbooks as fallacies inside language, or linguistic fallacies, for example, by Copi (1982). Yet this would appear to be an inappropriate classification, given that arguments from parts to wholes and wholes to parts do not appear to be linguistic fallacies in the same way that equivocation, accent, and amphiboly are.

The solution to this puzzle, already mentioned in chapter 2, section 13, above, is that Aristotle, in the *De sophisticis elenchis*, viewed composition and division as linguistic fallacies that had to do with the groups of words. As noted in chapter 2, Aristotle wrote (*De sophisticis elenchis*, 166 a 22) that the expression 'A man can walk while sitting.' is true in a divided sense but false in a combined (composed) sense, implying that he can walk-while-sitting. The fallacy here is a linguistic one that is closely related to, or even a subspecies of, amphiboly.

Perhaps what led to this confusion was that Aristotle, in the *Rhetorica* (1401 a 24–1401 b 30), used other examples of composition and division that have to do with physical parts and wholes, mixed in with his predominantly linguistic types of cases. Given the emphasis on the modern interpretation of composition and division in the standard textbook treatment, however, it would be better not to classify them as linguistic fallacies in the same category as equivocation and amphiboly. What is proposed here is that they should be treated, in part, as formal fallacies that have to do with structural relationships between parts and wholes, since arguments from parts to wholes and conversely have a formal structure, in the same way that arguments from 'all' to 'some' have a formal structure. To this extent, they are formal fallacies. But in part, they are informal fallacies that require reference to the specific features of a given case.

In Woods and Walton (1989) the fallacies of composition and division are given a formal analysis, within the formal theory of aggregates of Burge (1977). Unlike sets, aggregates are physical entities in space-time and are capable of change and going out of existence. Other properties of aggregates are explained in Woods and Walton (1989, 108). For the vast bulk of aggregates, some properties compose and divide in them, while others do not.

How does this theory apply to the kinds of cases of composition and division treated by the textbooks? Let us take a case in point.

Case 59

> All the parts of this chair are brown.
> Therefore this chair is brown.

According to the analysis of Rowe (1962), this argument is valid, but according to that of Bar-Hillel (1964), it is only valid provided you add a meaning postulate in the form of an additional assumption: when all the parts of a chair are a certain color, the chair is that color. You can only know whether this additional assumption applies to the case in question, however, if you know something about the particulars of the given case. In this case, the property of being brown does compose, making the argument a nonfallacious instance of composition. But you could not know that in advance, without knowing that the argument is about the colors of chairs and parts of chairs.

According to this theory, composition and division arguments are not valid or fallacious per se or in every case. They are nonfallacious where the property in question does compose or divide in the aggregate in question. But this condition can be established in a given case only by seeing whether an additional assumption applies to that case or not. A composition or division argument is fallacious, in a case like 57 above or in others featured in the standard treatment, where this assumption is not met by the case (even though it may appear to be initially).

This analysis applies primarily to physical parts and wholes, of the kinds featured in the modern interpretation of composition and division. It could also be applied to some cases of linguistic parts and wholes as well, however. But on this analysis, the fallacies of composition and division should not be classified as linguistic fallacies. Although they have a linguistic aspect, it makes most sense to classify them as formal fallacies, at least to the degree that consequent and the other fallacies in the chapter have been categorized as formal fallacies.

Whatever theory of the part-whole relationship is chosen as the theory to model arguments from parts to wholes and conversely, whether it be mereology (a part-whole logic) or the theory of aggregates, it will be a formal theory, like set theory or first-order logic. True, to apply such a theory to individual cases, additional assumptions will have to be made, which will vary from one type of case to another. Even so, given the primary nature of the formal aspects of judging such cases as fallacious or not, composition and division arguments should be categorized as formal fallacies, just as quantifier-scope fallacies, propositional fallacies like consequent, and syllogistic fallacies are called formal fallacies. Of course, what is at issue is whether these fallacies really are formal fallacies in some sense. And the same reservations attach to composition and division. The features of the individual property in a given case will determine whether or not that property is composable or divisible.

At any rate, we should cease treating composition and division as linguistic fallacies in the same category as equivocation (see chapter 2 on linguistic fallacies).

7. Inductive Fallacies

What evidently accounted for the traditional categorization of the kinds of erroneous inferences studied so far in this chapter as formal fallacies was that each, at least partly, involved some sort of deductive reasoning that can be modeled in some formal structure. The concept of deduction is important here because it contrasts with the informal fallacies studied in the previous chapter. What they seemed so often to involve was a presumptive structure of reasoning. 'Presumptive' here means subject to exceptions, in a way that contrasts with strict deductive reasoning.

But what about inductive fallacies relating to the various kinds of errors of sampling, polling, and probabilistic and statistical reasoning generally? These are not usually treated as formal fallacies, and they are certainly not deductive in nature. But they do have a structure of a sort, and they also appear to be different from the context-dependent, presumptive kinds of errors of reasoning typified as informal fallacies in chapter 2.

A number of the fallacies are analyzed in Walton (1989, chap. 8), but in that treatment it is emphasized that these incorrect arguments are often more in the nature of blunders than deceptive tactics designed to trip up an opponent in argument. Hence there it was found appropriate to speak of the error of meaningless statistics, the error of unknowable statistics, the error of insufficient statistics, and the error of biased statistics, as opposed to calling them fallacies. Post hoc is also partly an inductive error because it involves correlation, but in Walton (1989) this fallacy was portrayed as a presumptive fallacy on the grounds that causation is best seen as a practical and presumptive relation as opposed to an inductive one. We already noted in chapter 2, section 7, that presumptive and inductive arguments are often confused, and that the fallacy category of hasty generalization and its cognates are often lumped together in a confusing way.

Although it is difficult to draw these lines at any level of theoretical precision not open to philosophical disputation, we need practically to differentiate between inductive standards of argument and presumptive standards. Presumptive arguments are not based on polls, samples, statistical data, or other types of empirical evidence characteristic of inductive arguments. They are based merely on assumptions to which the other party in dialogue agreed so that the dialogue could go ahead provisionally. Their burden of proof is not positive in the way that is characteristic of deductive and inductive argumentation. A presumption doesn't have to be proved by positive evidence to make it acceptable. It has only to be free of refutation by evidence brought forward by an opponent to make it acceptable as argumentation as a basis for drawing tentative conclusions.

This is a fundamental point for the analysis of fallacies. If there are different types of arguments, and different standards of success or correctness (normative rules) for each type, then whether an argument is fallacious or not must depend on the prior question of what kind of argument it was supposed to be. This point appears to be often overlooked, but it has significant implications for any attempt to found a doctrine of formal fallacies. If an argument is deductively invalid, or has the form of a deductively invalid argument, it doesn't necessarily follow that it is a fallacy. For it could be an argument that is correct as an inductive argument and that, furthermore, was rightly put forward as an inductive argument by its proponent in a dialogue. This point has profound implications for any analysis of formal fallacies.

The subject of inductive fallacies contains a recent controversy that should be noted here. In investigating kinds of reasoning made by human subjects, experimental investigators of inductive fallacies like Tversky and Kahneman (1971) have assumed that the standard norms of statistical analysis, like the probability calculus, set the standard for a correct argument, so that deviations from this standard by experimental subjects can be called fallacies. But according to Cohen (1977, 261), this is a mistake that assumes unjustifiably that the subjects use the same concept of probability that the experimenters identify with the principles of the mathematical calculus.

If a coin is flipped, and keeps coming up 'heads' for a large run of tosses—for example, twenty—then experimental subjects will think that the odds of its coming up 'tails' on the next toss are higher than 50 percent. Strictly speaking, however, according to the standard rules of probability, this is a fallacy, at least on the standard assumption that each toss is independent of the next one. On the assumption of independence of each toss, the probability of heads on each individual toss is exactly 50 percent. This error is called the gambler's fallacy.

According to Cohen (1982, 260–63), however, it is possible for a person to interpret the notion of probability as intensity of belief, or strength of natural propensity, and these ideas do not conform to the principles of the mathematical calculus of probabilities. According to this interpretation, the gambler's fallacy need not be a fallacy. Cohen (1982, 262) concludes that "for the most part it is not a fallacy at all." We might note here that a very reasonable reaction after so many heads would be to doubt the assumption that the coin is untampered with. Erik Krabbe commented that he would put his stakes on yet another head to follow!

Without going into these matters in depth here, suffice it to say that a problem exists for inductive fallacies that is comparable to the problem we have found with the deductive, formal fallacies. Just the fact that the given argument has a form that is invalid, or does not conform to the requirements of the mathematical structure of reasoning, is not enough to condemn the argument as a fallacy (the conventional treatment notwithstanding).

8. Types of Arguments

One requirement of calling an argument an instance of one of the formal fallacies so far examined is that there should be some indication of what type of argument it is supposed to be—deductive, inductive, or presumptive. Whether an argument is an instance of a fallacy in fact often turns on this point. If it was supposed only to be a presumptive argument, but someone treats it as a very strict type of argument that should meet the requirement of deductive validity, then that mistreatment itself could be a "fallacy" of some sort.

One indicator of the type of argument in a given case is very often the warrant expressed by the major premise. Is it clearly meant to be a strict conditional, of the form, 'For all x, if x has property F then x has property G'? Or is it meant only to be a rough or presumptive generalization that can be maintained even in the face of some counterexamples (exceptions)? Sometimes it is clear from the context of dialogue how an argument is meant to be taken. And this needs to be taken into account in judging it as a fallacy or not.

One way of making sense out of so-called formal fallacies like affirming the consequent is to distinguish between the two kinds of conditionals that often function as warrants for drawing conclusions in arguments. One is the *strict conditional*, which expresses a conditional 'If A then B' that implies that it is impossible for A to be true and B false. The other is the *presumptive conditional*, which expresses a conditional 'If A then B' that allows it to be possible for A to be true and B false but implies only that it would be an unexpected (nonnormal) situation for this to obtain. With the strict conditional, if A is true, then B has to be true. You are inconsistent if you accept A and 'if A then B' but do not accept B. With the presumptive conditional, however, you can maintain such a stance provided you can give some evidence that while 'If A then B' is generally true, this situation is an exceptional one where this conditional defaults. This distinction is meant to be exclusive but not exhaustive of all types of conditionals. For example, there may be inductive conditionals that are neither strict nor presumptive conditionals.

Given this distinction, it is possible to argue that arguments having the form of affirming the consequent are not always fallacious. A good pair of examples that can be used to illustrate these two types of arguments is given by Sanford (1989, 40). Sanford calls the first case an "outrageous example," saying "no one would buy this argument."

Case 60

If a bolt of lightning kills you tomorrow, you won't live to be 125 years old.
You won't live to be 125 years old.
Therefore a bolt of lightning will kill you tomorrow.

This argument has the form of affirming the consequent. It is a kind of argument that the textbooks might use to illustrate the fallacy of affirming the consequent.

As our previous discussions of this type of case have shown, however, it would be a logical leap to call this argument an instance of the fallacy of affirming the consequent simply because it has this invalid form of argument, without addressing additional linguistic questions of whether this form is the explicit form, and so forth. Even so, this argument is, as Sanford says, "outrageous." Why do we perceive it so?

The likely reason, we suggest, lies in the plausibility of interpreting the conditional in the first premise as the strict conditional: for all x, if x is killed then x is dead (without exception, or even the logical possibility of such). And presumably, the individual in this case is less than 125 years minus one day in age. On such an interpretation, it is impossible for the antecedent to be true and the conclusion false. And affirming the consequent is (at least plausibly) a fallacy here. For in case 60, it is clearly false that it is impossible for the premises to be true and the conclusion false. The premises are (quite plausibly) true, while the conclusion is (quite plausibly) false.

Now consider the other example, which Sanford says "not everyone would immediately reject." (40).

> *Case 61*
>
> If he is a Communist sympathizer, he disapproves of our policy in Central America.
> He does disapprove of our policy in Central America.
> Therefore he is a Communist sympathizer.

As Sanford notes (40), this argument has "exactly the same form" as the preceding one. Is it therefore also a case of the fallacy of affirming the consequent? The suggestion put forward here is that it need not be and that it depends on how you interpret the conditional (and with that the concept of validity or warranted inference that supposedly links the premises to the conclusion).

Should we interpret the conditional in this case as a strict conditional? We might, but that interpretation would not normally (given no other information about the case) be the most plausible one. The more normal (and charitable) interpretation would be to say that from what we generally know of the Communist position, as expressed in the Communists' commitments and policies, we could generally infer with good (but not conclusive) justification that if x is a Communist, x would disapprove of our policy in Central America. (We presume here, in the absence of more definite information, that some features of this policy would run antithetical to the typical Communist position on such matters.) Evidence could be given to support this interpretation, perhaps based on what Communists have said in the past, or on some particular propositions known to be central to their position, although it would not be conclusive evidence.

When the case is so conceived, a body of evidence in case 61 leads us to presume that as a consequence of his being a Communist sympathizer he would likely disapprove of this particular policy. If this evidence is taken as true, then it could be presumed to set up a sign relationship between the antecedent and consequent. That is, it creates a presumption that disapproving of this particular policy could be an identifying sign or signaling indication of someone's being a Communist sympathizer. This could be wrong in some cases and could even be an innuendo, used to discredit persons unfairly labeled "Communist sympathizers." But in some instances, surely it could be a legitimate presumptive argument, used to make an allegation. Regarded as a presumptive argument, it could be reasonable (in some cases). In any event, it appears too strong, given the possibility of such an interpretation, simply to call the argument fallacious as such.

So in this case, the argument has the form of affirming the consequent, but for that reason alone, it should not be called a fallacy. It is a presumptive argument, one that requires further evidence to make it an argument with any rational basis for our acceptance in any given case under consideration. But should that factor, by itself, be enough for us to conclude that it is fallacious? It would seem best, on balance, to say no.

What's wrong, at least potentially, with this particular argument is that it could be an ad hominem attack[11] used to label someone unfairly as a "Communist sympathizer" and thereby raise a cloud of suspicion against anything he might further say as a credible spokesperson on foreign policy. But can we even go this far? The best answer, at least to this point in our knowledge of the ad hominem fallacy, is that we need to know more about the specifics of the case to legitimize such an interpretation. As things stand, the argument in case 61 could be legitimate, for all we know, even though its ad hominem nature certainly raises some legitimate suspicions about its use in the case in question. The context of dialogue gives insufficient evidence for us to nail down the argument as a fallacy.

Surprising as it seems, virtually every case studied in this chapter has tended toward refutation of the idea that so-called formal fallacies can be analyzed as fallacies because they are violations of the rules of some formal logical system. In every case, the so-called formal fallacy has turned out to be, or at least to involve, an informal fallacy. Or at least it cannot be straightforwardly and exclusively explained as a formally invalid argument of some type.

9. Fallacies and Logical Form

Can we really say that cases such as 44, 46, and 47 are formal fallacies, that the arguments in these cases are fallacious because they have a logical form that is invalid in propositional logic? We have found basically two objections to this traditional way of proceeding.

One is Massey's objection that just because an argument has a form that is deductively invalid, it does not follow that the argument is itself deductively invalid. To draw such a conclusion, you need to know, in addition, whether that form is the *maximally explicit form* of the argument, that is, whether the logical structure of the given argument has been fully revealed by the form.

In cases 44, 46, and 47, the form presented does seem to represent the explicit form of the argument, because every logical constant, every 'if . . . then,' 'and,' 'not,' and so forth, appears to have translated into the logical symbolism of the form. But how can we be sure that this is so, in a given case? It appears to be a linguistic question of how well the natural language argumentation in a given case has been rendered into the logical symbolism.

Moreover, questions also arise about the system of formalization that has been used. For example, modus ponens may be the explicit logical form of an argument as far as propositional logic goes, but what about quantifiers, modal operators, or other aspects of logical form that may not be rendered in a given symbolism? For we need to remember that there are different structures of formal logic that could be applied in some cases. It seems, then, that so-called formal fallacies are not purely formal but have a strong linguistic component as well.

The other objection we found is that an argument that commits a formal fallacy like affirming the consequent is fallacious, at least partly, by virtue of its resemblance to a valid form like modus ponens. This is the element of seeming validity, whereby a fallacy was traditionally said to be an invalid argument that seems valid. How important is this? Some would say that it is not important, because it is only a psychological question or a tangential matter of appearances. According to this viewpoint, what is primarily important is the normative question of whether the argument in question is correct or not, in relation to some normative model that defines what a correct argument is. With formal fallacies, the model is deductive logic of some sort, like propositional logic or syllogistic logic. With cases like 44, 46, and 47, for example, what makes them fallacious is that they have an invalid logical form in propositional logic.

This approach may work well enough for cases 44, 46, and 47, for these cases are transparently bad arguments, even to someone not familiar with formal propositional logic. A little reflection could show how to refute those arguments, using the method of refutation by counterexample. There is little danger that these arguments will be seriously deceptive to a moderately attentive audience. Indeed, they are really textbook examples of bad arguments, and they seem best suited to fulfill that role.

Aristotle's examples of the fallacy of consequent are different in this respect. They also have a form that is invalid in propositional calculus, but taken together, they show very well how the fallacy of consequent really works as a seriously deceptive kind of argumentation. When we get beyond the simplistic examples given in the standard treatment of the textbooks, we start to get a much richer analysis and explanation of how this fallacy works.

This observation suggests that the element of seeming validity may be not an accidental frill of the concept of fallacy or a purely psychological matter of how any particular individual or group responds to an argument that is fallacious. It suggests that the element of deceptiveness may be essential to the concept of fallacy and may be necessary to distinguish adequately between arguments that are fallacious and those that are merely invalid or are insufficiently supported by their premises. According to this viewpoint even the so-called formal fallacies like affirming the consequent are not purely formal in nature as fallacies.

Why Aristotle's examples of the honey and gall, the drenched earth, and the adulterer are fallacies remains something of a mystery. He seems to be right that it has something to do with the switching around of the conditional. They do seem to involve deductive logic or necessary inference, but they also seem to involve argumentation from sign.

The adulterer case is particularly interesting because it does seem to be a very common kind of argumentation that can easily mislead or be used to attack an opponent by innuendo and throwing a cloud of suspicion unfairly—effective and dangerous tactics in argumentation. It is a consequence of someone's being an adulterer that he wanders abroad at night, because (let's say in the spirit of Aristotle's example) that wandering abroad at night is part of the adulterer's normal means of carrying out his acts of adultery. Thus wandering abroad at night (N) is a sign of adultery (A), not a conclusive sign, but a small bit of evidence that would count against the case of someone against whom there is already a suspicion of being an adulterer. So we can say, in a given case, 'If A then N' is generally or normally true (but not true without exception). Can we then turn this statement around and conclude, 'If N then A.'? Yes, possibly we can, but the argument is even weaker this way, because there could be many other explanations of N that would not entail A. Hence turning the conditional around is dangerous and is apt to lead to an exaggerated assessment of the strength of the argument. As Aristotle notes, lots of people wander abroad at night who are not adulterers. But since Aristotle's case of the adulterer seems so clearly to be a presumptive fallacy of shifting the burden of proof based on suspicions, it may not even seem to be a formal fallacy at all. Is it really the same fallacy as the textbook case 46 of the fallacy of consequent?

In case 46, it would perhaps be normal to infer that if he went to Paris, he must have won the sweepstakes, given the assumption that he is not a wealthy person who could otherwise afford to go to Paris or would be likely to do so. But strictly, that is not what the premise says. It says, "He said he would go to Paris if he won a prize." It does not say that he would not go to Paris unless he won a prize. Hence, strictly speaking, to conclude that he won a prize, on the basis of the premise that he went to Paris, is not justified (by deductive reasoning alone). Hence the fallacy here, as in Aristotle's case of the adulterer, is one of inferring a presumption that may be true but may also be false.

So the two cases are somewhat alike and do both appear to involve the turning around of a conditional that is weaker one way than the other. But where Aristotle's example is much more complicated, and also much more interesting, is in showing how this kind of reasoning is so characteristically used to drive along suspicions based on guesses, hints, and presumptions as a case is gradually built up (or torn down) by an incremental growth in strength of evidence in a whole series of such signs that fit into a larger picture.

But this is a typical informal fallacy that works by the shifting of a burden of proof in presumptive argumentation. Curiously, then, the most mundane and ordinary formal fallacy of affirming the consequent turns out to be a lot more complex and interesting than any simple analysis of it as an invalid inference in propositional logic could ever reveal.

10. Fallacies as Failures of Use

One lesson this chapter has brought home is that fallacies—even those designated formal fallacies—are best seen as failures in the use of argumentation. Just because an argument has a form that is deductively invalid in some system of formal logic, for example, it does not follow that this argument is a fallacy. For it may never have been meant to be deductively invalid in the first place. Maybe it was really meant to be an inductive or presumptive argument of some sort that would be deductively invalid even when it is perfectly correct (according to the appropriate inductive or presumptive standards).

More characteristically, we have found, a fallacy is an argument that is used in such a way that it is pressed forward as appearing to be, or posing as, an argument that is more conclusive than it really is (when examined with more care). This seems to be a really pervasive characteristic of fallacies. They are arguments that in principle have some degree of correctness but are pressed forward in a given case too aggressively, or in an unwarranted fashion, masquerading as a much more powerful type of argument.

Thus it is never enough, in analyzing a fallacy, to show that it is an instance of some formally invalid type of argument. Such a normative failure is not sufficient, in itself, to characterize an argument as a fallacy. This same lesson could turn out to be true for informal fallacies as well. If there are various types of dialogue other than the critical discussion, it will not be enough to say an argument is a fallacy just because it violates a rule of a critical discussion. For although it may violate a rule of a critical discussion, it may be a perfectly legitimate (non-rule-violating) argument in some other context of dialogue, like a negotiation. What needs to be shown, then, is not only that the argument violates a rule of a dialogue of type x but that it was rightly supposed to have to meet the requirements of a dialogue of type x when it was advanced, in the given case.

It seems that fallacies often work because they are shifts, or masked duplicities, between one context of dialogue and another. This could perhaps be the basis of the old idea that a fallacy is an argument that seems valid but is not.

In chapter 5, it will be shown that there are different types of presumptive arguments, and each of these types has a characteristic argumentation scheme that displays the requirements for its correct use in a dialogue. Thus the argumentation schemes for presumptive reasoning will be seen to be comparable to the forms defining correct deductive reasoning like modus ponens. Perhaps inductive arguments are also correct or incorrect by virtue of forms or normative standards set for these types of arguments, modeled, for example, in the probability calculus or in accepted procedures for sampling, polling, collecting statistical data, and so forth. This development will put formal and informal fallacies on a more equal footing. For the biggest obstacle to analyzing informal fallacies has always been the lack of a clear or definitive account of these argumentation schemes.

4

Types of Dialogue

In order to evaluate an argument as correct or incorrect, it is vital to know the context of the conversation in which this argument was used. There are certain standard contexts in which arguments are typically put forward. To represent some of the most important and typical of these commonplace contexts, six normative models of dialogue are outlined below.

We have chosen the word 'dialogue' here, but the word 'conversation' is also appropriate. A dialogue is a conventionalized framework of goal-directed activity in which two participants interact verbally by taking turns to perform speech acts. Typically, these speech acts are questions and replies to questions. The various speech acts are linked together in a sequence that has a purpose and direction as the dialogue proceeds. The purpose is determined by the goal of the dialogue as a recognized type of social activity.

Each type of dialogue represents a context or setting in which argumentation occurs in everyday conversations. It is also important for informal logic to study dialogue contexts for explanation, description, and other types of discourse. But argumentation is our central focus here, and we restrict the treatment here to contexts of argumentation. The contexts identified will be defined as structures in which an argument is embedded, as used in a wider passage of discourse.

These structures are called *normative models*, meaning that they stipulate how an argument should go as an ideal conversational exchange where two parties reason together for some common purpose.

The six models presented are not, of course, the only types of dialogue in everyday communication.[1] But they represent six of the most important types, from the viewpoint of evaluating argumentation. And many other recognizable types of dialogues can be shown to be mixed types that can be shown to be compounds of two or more of these basic types.[2]

It is important to recognize that the primary purpose of these models of dialogue is not to describe, psychologically or empirically, how people actually argue in everyday conversations. Instead, they are supposed to represent how people ought ideally to argue if they are being "reasonable" in the sense of adhering to collaborative maxims of politeness that enable a conversation to go ahead in a productive manner.

The underlying principle is the Cooperative Principle of Grice (1975, 67): "Make your conversational contribution such as is required, at the stage at which it occurs, by the accepted purpose or direction of the talk exchange in which you are engaged." As Grice emphasizes (69), standard types of talk exchanges (dialogues) can be seen not only as empirical descriptions but as normative models that define practices giving a standard of "something which it is *reasonable* for us to follow" as opposed to something most of us do in fact follow.

From the viewpoint of the analysis of fallacies and other kinds of critical shortcomings studied in informal logic, the critical discussion would appear to be the most important or central type of dialogue. These other types of dialogue can be viewed as clustering around the critical discussion. Certainly the critical discussion has been the most thoroughly analyzed model in the literature on argumentation—see van Eemeren and Grootendorst (1984; 1992).

But often, with fallacies and other critical errors, the underlying problem is a subtle, undetected shift from one type of dialogue to another—often it is from a critical discussion to some other type of dialogue. So these other types of dialogue turn out to be very important as well in understanding and evaluating cases of fallacies.

1. The Critical Discussion

The goal of the critical discussion as a type of dialogue is to resolve a conflict of opinions. What is meant by 'resolve' is more than just to end the conflict but to end it by some means of reasonable argumentation, so that the one opinion is seen to be better supported by the evidence than the other. Van Eemeren and Grootendorst (1984) describe in detail the four stages of a critical discussion—the opening stage, the confrontation stage, the argumentation stage, and the closing stage. The rules given by van Eemeren and Grootendorst (1984; 1992) define the kinds of arguments that are acceptable at any given stage of a dialogue.

According to van Eemeren and Grootendorst (1984, 90), there are two basic types of critical discussion. In the *simple critical discussion*, one participant defends a particular proposition known as her *thesis*, and the other participant has the role of raising critical questions that cast doubt on that thesis. In the *complex critical discussion*, each participant has a thesis, and the goal of each participant is to prove that his or her thesis is true. The complex type of critical discussion, also called a *dispute* in Walton (1989b, 286), is a *symmetrical* type of dialogue in the sense that both participants have the same kind of task or role in the dialogue.

A more general type of dialogue, of which the critical discussion can be classified as a subspecies, is called the *persuasion dialogue* in Walton (1989a, 5–6). In a persuasion dialogue, each participant has the goal of persuading the other participant that her (the first participant's) thesis is true (or at least acceptable, on balance of considerations). The way for a participant to go about this task of persuading in such a dialogue is to advance arguments that have the other party's commitments as premises and (ultimately) one's own thesis as a conclusion. What is distinctive about persuasion dialogue is that in order to prove anything successfully, we must derive it by acceptable arguments from premises that *the other party* is committed to.[3] In other words, argumentation in a critical discussion is, by its nature, directed toward the other party and is based on that other party's commitments. We must always ask: what will successfully persuade this particular person (or audience)?

The concept of commitment (Hamblin 1970; 1971) is fundamental to the structure of persuasion dialogue. The idea is that (ideally) each participant has a repository, a kind of data bank that keeps track of all the propositions that he or she has become committed to, at any given stage in the sequence of dialogue. Commitment is not a psychological concept for Hamblin (or for van Eemeren and Grootendorst either). It is a normative concept. Your commitments are the propositions that you have (explicitly or implicitly) inserted into your commitment store by virtue of a certain type of move that you have made in a certain type of dialogue.

Persuasion dialogue is a general or generic type of dialogue, defined only in terms of commitments to be used in argumentation. The critical discussion is a much more specific and precisely regulated type of dialogue that has all kinds of specific rules defining what a participant may or may not do at any given stage.

The collective goal of the critical discussion as a type of dialogue needs to be carefully distinguished from the individual goals of the participants.[4] The goal of the critical discussion generally is to resolve a conflict of opinion by rational means. But the goal of each participant individually is to prove that his or her point of view is right. A *point of view* (van Eemeren and Grootendorst 1984, 89) is a proposition, taken together with an attitude (pro or contra) with respect to that proposition. The attitude of *critical doubt* is not the same thing as a contra attitude but instead is a suspended point of view that shifts a burden of proof onto the other party through the asking of certain types of legitimate and appropriate critical questions. In general a participant succeeds in proving his thesis by fulfilling a *burden of proof*, a weight or preponderance of evidence that is sufficient to constitute proof of a proposition.[5] The burden of proof is set (ideally) during the opening and confrontation stages of the critical discussion.

In a critical discussion as a model of rational argumentation, it is important and indeed crucial to keep track of the commitments of each participant as the dialogue proceeds. For this purpose, a list or store of statements, called a commitment store in Hamblin (1970; 1971), should be kept. Generally, the participants will begin a persuasion dialogue with some initial commitments, and then as they put forward speech acts (or moves) in the dialogue, statements will be inserted into, or deleted from, their commitment stores. For example, when a participant makes a speech act of assertion, "I assert proposition *A*," then *A* goes into her commitment store. Or when a participant makes a retraction, saying "I am no longer committed to *A*," then *A* is removed from his commitment store.

For Hamblin, the commitment stores were generally represented as propositions that are clearly on view to the participants in a dialogue. This feature represents one type or level of an ideal, rational argument exchange where the participants always remember what they (and the other parties) have committed themselves to by their past utterances in a dialogue.

Realistic argumentative exchanges in everyday conversations, however, frequently do not reach this level of rationality or cooperativeness. Sometimes participants forget what was said previously, and sometimes they even deliberately lie about what they said or were committed to by their speech acts in the past. Ideally, a record (a tape or transcript) ought to be kept to resolve such questions. But in reality there may be no record of this sort. Sometimes, also, one's arguments are expressed in a vague or ambiguous, or simply unclear way that leaves the question of whether a proposition is a commitment or not subject to interpretation. For this reason, the distinction was made (Walton 1984) between explicit and nonexplicit commitments.

Some commitments are *explicit (light-side) commitments*, meaning that they have been explicitly conceded as propositions accepted by a participant, through some speech act she has made in the dialogue, or have been explicitly laid down as a commitment during the initial stages as initial commitments. These are called light-side commitments in the sense that they are known to the participants and are on view—you can see them by simply looking into the participant's commitment store. Such a set could be a list of propositions on a blackboard, for example, or in a computer memory.

Other commitments are *nonexplicit (dark-side) commitments*, meaning that they have not been explicitly conceded but can only be conjectured, or inferred obliquely from what is known, from the underlying position of a participant, as expressed in his arguments in the dialogue.[6] These are called dark-side (veiled) commitments in the sense that you can't actually see them on view by looking directly into a participant's commitment store. They do exist in that store, or they don't, but you can't find out by simply looking or checking the store.

To try to confirm whether someone is definitely committed to a particular position by getting it from their dark-side set of commitments to their light-side set, you have to draw out plausible inferences and ask questions. For example, suppose George has always, in the past discussion, been committed to socialism and a left point of view in politics, but then advocates that the post office be run by free enterprise. Here his dark-side commitment to socialism appears to conflict with his commitment, in this particular argument, to something that seems to go against socialism. A questioner could ask him: "Are you serious, George? I thought you were a committed socialist. How can you resolve this apparent conflict?" Now, perhaps George can resolve the conflict. Perhaps he is a modified socialist of some sort or can explain how a free enterprise post office is compatible with his brand of socialism. But by using his dark-side commitment, the questioner can shift a burden of proof onto George's side of the dialogue to resolve the apparent conflict in order to defend his point of view.

Van Eemeren and Grootendorst think of the goal of a critical discussion as the resolution of a conflict of opinions. But in many cases, a critical discussion can be very valuable and informative even though a definitive resolution of the conflict is not achieved. But once we have brought in the distinction between light- and dark-side commitments, an important benefit of such a critical discussion can be identified. This is the benefit (Walton 1989b; 1992c) of increased understanding of the argumentation behind the opponent's point of view and, perhaps even more significantly, the argumentation behind our own. Through a critical discussion of an issue, our point of view on it may become more deep, subtle, and strongly supported, even though the proposition we are defending at the end is the same one we defended at the beginning of the dialogue.

Although it is not the main goal of a critical discussion, a very important side benefit is the fulfillment of the *maieutic function* of bringing a participant's key underlying dark-side commitments to expression as light-side commitments.[7] This term comes from the Greek word *maieutikos*, meaning "skill in midwifery," and refers to the skill attributed to Socrates of being able to assist in the birth of new ideas (mental offspring) through philosophical dialogue with another party, who would express these ideas, with the help and questioning of Socrates. Thus the maieutic function of dialogue is the enabling of a participant to express her previously unarticulated but deeply felt commitments in a much more explicit and carefully qualified way through testing them out and trying to defend them in a reasoned discussion with another party who may be skeptical about them or not so inclined to accept them initially.

It is a good question whether this maieutic function leads to knowledge or only to a kind of insight or increased understanding of one's own personal views and commitments. But even if the latter is the only real gain, nevertheless that could be a very important kind of benefit or advance that could prepare the way for knowledge. The advance here could be described as a kind of negative clearing away of prejudices, bias, dogmatic preconceptions, fallacies, and so forth that removes important impediments to the advancement of knowledge.

2. The Negotiation

Negotiation dialogue is quite different from the critical discussion, because the goal of a participant is not to prove or argue that some proposition is true or false by marshaling evidence. By contrast, a participant in negotiation makes offers and concessions in order to "get the best deal." It is not truth but money (or some kind of goods or economic resources that can have financial value or implications) that is at stake.

The initial situation in negotiation dialogue is a set of some given goods or services that are in short supply, such that both participants cannot have all they want. The goal of a participant in this type of dialogue is to maximize his or her share of these goods or services by verbal means of securing agreements with the other party. This goal is achieved by a process of bargaining in which the strategy is directed to finding a compromise that will be acceptable to both parties. The goal of the negotiation dialogue is to "make a deal," to reach an agreement that both parties can live with even if it involves compromises. Both sides try to get what matters to them most and to trade off concessions that are less important or essential from their point of view. In a successful negotiation, the positions of both sides converge from extremes or opposites at the beginning, toward a middle position that is acceptable to both.

According to Walton and McKersie (1965), there are four subtypes of negotiation dialogue. In *distributive bargaining,* the activity most familiar to students of negotiation, the goal of the one party is in basic conflict with the goal of the other (4). To be in *basic conflict* in this sense means that the dialogue is a zero-sum game between the participants: "one person's gain is a loss to the other" (4). The *issue* of such a negotiation is the area of common concern to the participants "in which the objectives of the two parties are assumed to be in conflict" (5).

Integrative bargaining is a type of negotiation where there is no basic conflict between the goals of the participants. Instead, the area of common concern is a *problem* (5), where the interests of both parties can be integrated, to some degree.

In *attitudinal structuring,* the issue is not purely economic but concerns relationships between the participants, in particular attitudes like "friendliness—hostility, trust, respect," and "motivation—orientation of competitiveness-cooperativeness." This type of negotiation dialogue seems to have more to do with personalities than, at least directly, with money, or overtly economic considerations.[8]

Intraorganizational bargaining is a type of negotiation dialogue in which the goal is to bring the expectations of one side into alignment (5). For example, in labor negotiations, the local union and the international union may have to get together and agree on their objectives. On the company side, management and staff groups may have to get together and discuss their differing aspirations (6). This type of negotiation presupposes considerable broad agreement of objectives at the outset, for both parties belong to the same group and are on the same side, in a prior and broader context of negotiation.

Generally, the concept of commitment is very important in the analysis of negotiation dialogue given by Walton and McKersie (1965). They see commitment as "the act of pledging oneself to a course of action" (50). A commitment is a statement of intentions that may be a *threat,* in the sense that the "strategy selected will have adverse consequences" for the other party (50). This idea that making a threat is, at least in some instances, a legitimate part of the argumentation in a negotiation dialogue, is very important when it comes to the study of fallacies.[9] According to Donohue (1981a, 279), promises and threats are among the core concepts of bargaining theory. Threats are among the legitimate tactics of negotiation listed by Donohue (1981a), but they are also said to be "among the most high risk tactics," because "they are often viewed as a final stand" (279). *Tactics* in negotiation are techniques of argumentation that can be used by the participants to achieve their goals successfully.

Donohue (1981b) analyzes negotiation as a normative model of argumentation by setting out rules that define good or successful negotiations. *Constitutive rules* define "how we are to interpret the sequence of utterances" (108), while *regulative rules* "govern the prescriptive nature of the communication event," defining what "specific behavior is expected" and that deviations from the prescribed behavior are "subject to evaluation" (108). For example, if one negotiator attacks the position of the other, then the other is "likely to be under some intense prescriptive force" to respond (108). Failure to respond will result in the evaluation that the other concedes the point being attacked (109). According to Donohue (1981b, 109), prescriptiveness is not judged by a third party in negotiation dialogue but is monitored by the participants themselves.

The prescriptive rules for negotiation function by requiring the respondent to respond in a particular, prescribed way to the use of a given type of move or tactic by the other party. Failure to respond in the right way "can be viewed as tacitly conceding or supporting the point being attacked" (Donohue 1981b, 112). The use of an argumentation tactic in negotiation dialogue has the function of shifting an obligation, or burden to respond in a certain way, onto the respondent. Failure to respond by challenging the attack successfully and fulfilling the obligation is evaluated as conceding the proponent's argument. According to Donohue (1981b, 112), "failure to challenge can be viewed as support for the attacking points." Argumentation in negotiation shifts a burden, or obligation to respond, back and forth as the dialogue proceeds. How this shift works is determined by the type of move made and its place in the negotiation.

Generally, the main thing to be clear about with respect to negotiation is that it is a legitimate type of dialogue in its own right, in which argumentation may occur, even though the principal goal of the argumentation is not to discover the truth. If two parties are negotiating with each other, it is quite correct and accurate to describe what they are doing as argumentation. They are "arguing with each other," even though they are not trying to resolve a conflict of opinions or to show that some proposition advocated by the other party is false or not supported by sufficient evidence for its acceptance.

Negotiation is a legitimate context of dialogue in which argumentation can be evaluated as right or wrong, correct or incorrect, good or faulty, in relation to its contribution to the goal of that type of dialogue. Of course, it is quite another matter if the participants were supposed to be engaged in a critical discussion in the first place or in some other type of dialogue where the primary concern should be the discovery of truth or falsity of a proposition and then one or both parties covertly or illicitly shifted to negotiation or "making a deal" with respect to accepting or rejecting that proposition. In this kind of case of a shift, the argumentation must properly be judged from the normative point of view of the goals and rules of the first type of dialogue—the one the participants were supposed to be engaged in at the outset of their argumentation. Such a case is different from the kind of case where the participants were supposed to be negotiating in the first place and there was no shift.

3. The Inquiry

The goal of the inquiry is to prove whether a particular proposition is true (or false) or, alternatively, to show that, despite an exhaustive search uncovering all the available evidence, it cannot be proved that this proposition is true (or false). The initial situation of the inquiry is the problem posed by a need to establish, one way or the other, whether a particular proposition is true or not. For example, if there has been an air disaster, it may be very important, for reasons of air safety, and to satisfy the families of the victims, to settle lawsuits, and so forth, to try, insofar as it is possible, to determine exactly what happened, that is, what caused the disaster. In order to do this, an official government inquiry may be undertaken.

When it is said that the goal of the inquiry is to *prove* something, 'proof' is meant here in a way that implies a very high standard, or heavy burden of proof. 'Proved' in this sense, means definitely established on the basis of premises that are known to be true. This standard of proof implies that all the available, relevant evidence has been collected and carefully stated in such a way that none of it should need to be retracted in the future.

A most important characteristic of the inquiry as a type of dialogue is that it is meant to be *cumulative,* in the sense that the line of reasoning always moves forward from well-established premises to conclusions that are derived by very careful (ideally, deductively valid) inferences, so that the conclusions are solidly established.[10]

The design of cumulative argumentation is that there should be no need to go back to modify or retract one's previous conclusions, because doing so would disturb the whole structure on which the final conclusion was built. A cumulative structure of argumentation is often compared to a building constructed on "solid foundations." The cumulative type of argumentation is often called "foundationalism" in philosophy as a philosophical method or theory.

In practice, there are many different kinds of inquiries, like the coroner's inquiry, other kinds of legal inquiries, and government inquiries. Each has its own special methods and standards of proof. Typically, it seems, official government inquiries are launched when there is a perceived crisis or problem of public opinion or popular concern that such-and-such a problem needs to be thoroughly investigated. Experts are then called in to conduct and contribute to the inquiry. Inquiries of this sort are often very expensive.

Once an inquiry has been launched, the first part of the argumentation, or main stage, is the collecting of evidence. Scientists or experts may collect a lot of relevant data, and these people, along with other experts who have consulted, or who have examined the data, will then testify as to how to interpret these findings. The next stage is a discussion stage, where the parties to the inquiry try to agree on what conclusions can be drawn from the evidence. Finally, there is a presentation stage, where someone is designated to write up some sort of report or document that gives the results of the inquiry. The order of the reasoning, from premises to conclusion, is, in general, different in the presentation stage from that of the previous stages, where the data were collected and evaluated.

When the inquiry is presented to its external audience, aspects of pedagogy and persuasiveness are very important. The presentation is supposed to be orderly, so that everyone who needs to know about the results can appreciate that the process of inquiry was thorough, orderly, and exhaustive in searching out all the evidence and drawing careful conclusions from it.[11]

A very good formal model of the logic of the reasoning in the inquiry is the semantics for intuitionistic logic presented by Kripke (1965). This model is based on a tree structure that represents "evidential situations" as the nodes or points of time where propositions are "verified" or "not verified." Then as the inquiry progresses, we find "new knowledge," and we progress along a branch of the tree, reaching a new node where more propositions are verified. As we go along the tree, which branches toward the future, more and more propositions are verified as the inquiry progresses. But the structure is cumulative, in the sense that as we go toward the future, propositions are never dropped (retracted or "deverified").[12]

It is interesting to see that we never have circular argumentation in a Kripke structure of this type. We never "loop back" to the past. The process of reasoning always unfolds toward the future in an expanding tree structure as more and more propositions are established or are verified. This feature was used by Woods and Walton (1978) to provide a model of argumentation in which circular reasoning is not allowed.[13]

Certainly, one property that is very important in the inquiry as a type of argumentation is *evidential priority*, meaning that the premises are better established or are more reliable as evidence than the conclusion they were used to prove. To put it another way, the conclusion was more doubtful than the premises, so that the premises can be used to prove (remove doubt from) the conclusion but not vice versa. Contexts of argument where evidential priority is important tend to be inimical toward, or to exclude, circular argumentation from counting as an acceptable way of proving a conclusion.[14]

It is a perennially interesting question whether the kind of argumentation used in, or appropriate for, science is that of the inquiry. In their rhetoric, scientists have often been known to propound accounts of scientific reasoning that make it sound like an inquiry. Descartes is known as an exponent of this view, and during the heyday of logical positivism in the twentieth century, the view of science as a type of inquiry was very popular.

The way scientists actually argue and resolve their disputes, however, often seems altogether unlike an inquiry and perhaps more like a critical discussion. And certainly we have seen that, in practice, scientific results often have to be withdrawn or corrected, in some cases even retracted because of fraud and faked results. The currently popular opinion among the philosophers of science would seem to be that science is not like an inquiry, or at least not very much, and is more like a running dispute or critical discussion in which there are conflicts of opinion and opposed groups struggling to promote their point of view and refute those of their opponents. In discussing this question, one ought to separate carefully the empirical question of how scientists actually argue in their professional pursuits from the normative question of what form a good scientific argument should take as a type of reasoning. The first question is one for sociology and philosophy of science, whereas the second question is one for logicians or argumentation theorists.

No claim is made here, however, that science is either an inquiry or a critical discussion. It is quite enough for our purposes here to recognize that scientists, in their rhetoric, convey an ideal image of scientific argumentation as a kind of inquiry. While retraction of announced scientific results is sometimes necessary, in reality—see Broad and Wade (1982, 181–92)—the ideal of science when portrayed as a type of inquiry is to eliminate the need for retraction as far as this is possible.

Retraction is the fundamental problem of managing commitment in formal structures of dialogue. And the key difference between the inquiry and the critical discussion as types of dialogue is how retraction is made possible. The critical discussion should be an open and fluid type of dialogue where retraction is generally (but not always) allowed. But the whole aim of the inquiry is to prevent the need for retraction.

Euclidean geometry is a good example of cumulative reasoning in science, within the context of an inquiry. Evidential priority is applicable. A circular argument that went from a later-numbered theorem as premise to an earlier-numbered theorem (or to an axiom) as conclusion is clearly meant to be unacceptable as a proof. Mackenzie (1980) showed that circular reasoning is not meant to be tolerated as an acceptable type of argumentation in this context.

4. The Quarrel

In any lasting relationship between two parties, there will be perceived harms, slights, or grievances on both sides that will not be explicitly stated by the one party and will not be noticed by the other party. The reason that such complaints are so often not stated in conversations is that it is necessary for smooth functioning of social and business concerns that constructive agreement be stressed and that comparatively small differences or disagreements be hidden or shelved. Also, in the course of many types of conversation, dwelling on complaints of perceived slights would not be an accepted part of polite conversation. Such complaints are to be made, if at all, in an aggrieved outburst that is an interruption, a shift out of a polite conversation to a different type of discourse.

The goal of the quarrel as a type of dialogue is for these hidden grievances to be expressed explicitly, acknowledged and dealt with, in order to make possible the smooth continuance of a personal relationship. Thus the chief benefit of the quarrel is to achieve a cathartic effect whereby these hidden conflicts or antagonisms can be brought out into the open and acknowledged by both parties to a dialogue. The closing stage of the quarrel is the healing of the opening in the relationship caused by the revealing of this antagonism. The opening stage is where the antagonism is expressed overtly, at first usually primarily by one party.

The goal of each party in a quarrel—as opposed to the collective goal of the quarrel as a type of dialogue—is to hit out verbally at the other party. The quarrel arises out of a feeling of truculence or resentment at some hidden injury that gnaws at a person, under the surface. Once this feeling is first expressed, the beginning of a quarrel is like a volcano erupting—all the hidden feelings pour out.

The quarrel typically begins over some trivial dispute or sparking incident that may have nothing to do, really, with the grievances that underlie the quarrel. Once these hurt feelings pour out, they are often about things that are not particularly relevant to the initial dispute that provoked the quarrel in the first place. In other words, the quarrel sustains a high degree of irrelevance in argumentation. It often skips from one incident to another, apparently unrelated incident.

A successful quarrel does a good job of exposing these significant but buried hurt feelings during the argumentation stage, as each party becomes aware of, and sensitive to, these grievances. Typically one person says, "I didn't realize that was so important to you." During a good quarrel, the participants "make up" during the closing stage, vowing to be more thoughtful or sensitive about this particular issue in future conduct.

Infante and Wigley (1986) have studied the quarrel empirically by means of a verbal aggressiveness scale. They argue that verbal aggression is worthy of study in its own right as a type of discourse in speech communication. By specifying different types of verbally aggressive messages and indicators, they have given various signs to identify when "rational discourse" has shifted to verbal aggressiveness in argumentation. These include character attacks, competence attacks, insults, maledictions, ridicule, and profanity (61). The dialogue often begins as a critical discussion, and then the presence of these signs indicates a shift toward a quarrel. In such cases, the quarrel may begin to intrude gradually, with one or both parties being unaware of the shift.

The argumentation stage of the quarrel is typified by counterblaming arguments. Each party attacks the other party personally for some fault or alleged personal breach of standards of good conduct. The attacked party is said to be guilty of having committed some culpable action in the past, perhaps on repeated occasions. The argumentum ad hominem is strongly associated with the quarrel as a type of dialogue and is often the key sign that a quarrel has begun or is under way.

It seems strange at first to think of the quarrel as a normative model of dialogue in which argumentation can be judged as good, legitimate, or correct. For it has generally been assumed in the past that the quarrel is inherently bad as a type of dialogue—something to be avoided and condemned.[15] Certainly it is true that a quarrel generates more heat than light and that the quarrel is not a central paradigm of good, logical argumentation in the way that the critical discussion is. But even so, according to the point of view advocated here, the quarrel is a normative model of dialogue in its own right, and arguments can be good or successful in a quarrel, provided that they contribute to the goal of the quarrel.

Thus the somewhat novel thesis being argued for here is that the quarrel can be a good thing or at least that argumentation in a quarrel can be correct or successful within that context. It is still maintained, however, that the quarrel is often associated with fallacies, because there is often a shift from another type of dialogue, like a critical discussion, to a quarrel. To judge such an argument as fallacious, however, you need to evaluate it in relation to the original type of dialogue in which the participants were supposed to be engaging in the first place. If that was a critical discussion then the quarrelsome arguments need to be judged on the basis of whether they contribute to the goals of the critical discussion and follow the rules appropriate for a critical discussion. In general, quarreling is a very poor way of forwarding the goals of a critical discussion.[16] More often, it blocks the goals of a critical discussion. Hence fallacies are often associated with an illicit shift from a critical discussion to a quarrel.

The quarrel is a type of eristic dialogue (from the Greek word *eris* for 'strife') that is dominantly adversarial and noncollaborative in nature. Eristic dialogue could be described as a kind of verbal combat where each party tries to win, at all costs, in order to humiliate and defeat the opponent. Even eristic dialogue is not purely adversarial or anarchical, however. The participants do take turns. But they use aggressive, unfair, and fallacious arguments whenever they think they can get away with it.

The quarrel is typically an emotional type of dialogue that "erupts," rather than being deliberately started by the participants. One particular type of eristic dialogue, however, is more of a deliberate intellectual exercise, designed to impress onlookers that an arguer is very clever and knowledgeable. This type of eristic dialogue could be called *sophistical dialogue*—it is a kind of staged intellectual quarrel where the participants use clever arguments to try to defeat each other in order to impress a third-party audience with their intellectual prowess.

The classic case of this type of dialogue is the part in Plato's *Euthydemus* that portrays the clever sophists attacking each other with all kinds of tricky arguments and clever verbal traps. Aristotle, in his *On Sophistical Refutations*, described eristic dialogue as a kind of "contentious reasoning" that is like cheating in sports, or unfair fighting, to win a victory at any cost (1955, 171 b 24–171 b 30):

> Just as unfairness in an athletic contest takes a definite form and is an unfair kind of fighting, so contentious reasoning is an unfair kind of fighting in argument; for in the former case those who are bent on victory at all costs stick at nothing, so too in the latter case do contentious arguers. Those, then, who behave like this merely to win a victory, are generally regarded as contentious and quarrelsome, while those who do so to win a reputation which will help them to make money are regarded as sophistical.

Following Plato, Aristotle condemned the sophists as a professional class of arguers who used their skills of eristic dialogue to make a profit. Calling the sophists professional quarrelers and suggesting that they were dishonest or biased because they took fees for their lectures was probably not very fair to the sophists, from what we know about them. Not many of their writings survived, however, and the condemnation of the sophists by Plato and Aristotle not only left them with a bad reputation but tended to discredit any kind of opinion-based argumentation, whether quarrelsome or not, for subsequent generations. At any rate, it is certainly right to say that the quarrel is not much of a friend of logic and that when another type of dialogue shifts to a quarrel, it is generally a bad sign.

According to the classification proposed here, eristic dialogue is the most general category, and quarrelsome and sophistical dialogue are subtypes of eristic dialogue. Quarrelsome dialogue is that type of dialogue where the participants try to blame the other party for some wrong allegedly committed in the past. The aim is to humiliate or cast blame on the other party through a personal attack. In sophistical dialogue, the aim is to impress an audience (or third party) by showing how clever you are in attacking your opponent in a verbal exchange and showing how foolish her views are. Both subtypes are classified as eristic dialogues because the goal is to defeat the other party at all costs.

The eristic dialogue is unique as a type of dialogue, of all the types of dialogue studied here, because it is a zero-sum game, in the sense of being completely adversarial—one party wins if and only if the other party loses. All the other types of dialogue are based on the Gricean cooperativeness principle (Grice 1975) mentioned on page 99. But in the eristic type of dialogue, the general presumption is that a participant is flouting the cooperativeness principle.

Leeman (1991, 51), writing on the rhetoric of terrorism and counterterrorism, defines the totally adversarial attitude as the principle "If you are not for us, you are against us." This closed attitude is characteristic of what is called (Walton 1992c) the group quarrel—an institutionalized, systematic type of eristic dialogue.

Both Braden in his study of white supremacists and Scott in his essay on black militants found that these speakers primarily justified themselves by denouncing those they were against. Similarly, for terrorists the "violence of the system" bipolarly balances the "violence of the terrorists." The choice becomes one of "either-or," either *for* the terrorist or *against* the terrorist. MacDonald summarizes this position directly. "More and more it will be a case of either being for us, all the way, or against us." For the Tupamaros West Berlin, violence against the system was the only escape from the system. "[You] cannot be neutral. Otherwise, you yourselves will be destroyed. You yourselves must beat and rob these pigs, burn their palaces, fight your oppressors, or you yourselves will be destroyed." [Leeman, 1991, 51]

This observation shows why eristic dialogue does not rest on the co-operativeness principle, at least not in the same way or to the same extent that the other types of dialogues do.

On the other hand, the quarrel, as a normative model of dialogue, does require a certain minimal degree of cooperativeness. For example, to have a good or productive quarrel, it is necessary for the participants to take turns. Each party must allow the other room to respond to his or her arguments for the quarrel to be a bilateral exchange that has a direction and flow. The degree of cooperativeness required to sustain eristic dialogue is very minimal compared with the other types of dialogue, however. In a quarrel, victory at (almost) any cost is the goal, regardless of the worth of an argument in, say, a critical discussion.

For this reason the shift from any other of the types of dialogue to a quarrel is uniquely negative, from the point of view of fallacies and argumentation. Hence also for this reason fallaciousness is identified with quarreling and with a quarrelsome attitude, as stressed by Aristotle so often in the *De sophisticis elenchis*. Plato and Aristotle, champions of dialogue as a method of philosophical reasoning, were nevertheless very worried about the negative side of the method they called 'dialectic.' This negative side is the descent or degeneration of any other kind of dialogue into eristic dialogue.

5. Information-Seeking Dialogue

Information-seeking dialogue is based on an initial situation where one party has some information that the other party needs or wants to find out about. This type of dialogue is very definitely asymmetrical. The role of the one party is to give or transmit information that she possesses. The role of the other party is to receive, or gain access to, that information. In this respect, the information-seeking dialogue is different from the inquiry, where the participants are all more or less equally knowledgeable or ignorant and their collective goal is to prove something.

During the argumentation stage, the one party, the information seeker, asks questions of the other party, who could be called the source, the respondent, or the informant. In this type of dialogue, a "why" question is typically a request for an explanation of a particular proposition and not a request to prove or support it by argument, as it is in the critical discussion.[17] The questioner or information seeker also asks yes-no questions, and the respondent is obliged to give a direct answer where possible or, if not, to explain why the question cannot be answered directly.

The goal of information-seeking dialogue is the transfer of information from one party to the other. We could say that the one party is ignorant and the other party has information. The goal is to redress this unequal distribution of information, to aid in carrying out some purpose. An example of information-seeking dialogue would be the conducting of a recruiting interview by a representative of a company looking to hire a new employee. It has been recognized by experts in recruitment interviewing that forming a first impression of a candidate within the first few minutes is an error. Instead, the importance of asking good questions is stressed as a basis for getting information to judge a candidate's abilities. Good questions seek out the information relevant to ability, and bad questions can interfere with the dialogue (Gay 1992, 522).

> Is there such a thing as a bad interview question? Yes, says Jonathan Siegel, a psychologist with the executive search and assessment firm of Westcott Thomas & Associates.
>
> These questions judge the candidate from a bias outside the criteria prepared before the interview. Candidates quickly sense there is a right and wrong answer. The consequence is doctored responses or defensiveness, which are both counterproductive.
>
> Bad interview questions include personal questions, frequently used to judge a woman's commitment level, and those that require only a yes or no answer. Bad interview questions can also be generic, randomly cadged from interview manuals. If questions are not tied to the job or the candidate's experience, candidates can answer in a way they believe best satisfies the interviewer's needs.

Here the purpose is to hire the best candidate, and the dialogue is to find the information relevant for that purpose.

One important subtype of information-seeking dialogue is the expert consultation dialogue, where one party is an expert (the source), and the other party is a layperson (the questioner). The questioner has a specific problem or need for a certain kind of information, and the expert is supposed to give advice or explain things in a way that the layperson can understand. There are many communication problems inherent in this type of dialogue, for it is difficult for experts to explain matters in their field in a way that is clearly intelligible and useful for a layperson in that field. The traditional fallacy of the argumentum ad verecundiam, the argument from respect for authority, is a general term to cover breakdowns and failures of argumentation in this type of dialogue.

The respondent doesn't always have to be an expert, however. If one person walks up to another person on the street in Leiden and asks that other person where the Central Station is located, the informant is not necessarily an expert on Leiden streets, like a cartographer, for example, or an urban affairs specialist. But such a person could be a helpful source if he is familiar with Leiden, that is, if he is in a position to know how to reach the Central Station from the present location.

In this type of case, the information-seeking dialogue may be functionally related to a deliberation type of dialogue. The first person may be reasoning: "I need to get to Central Station. How should I do it? Should I go this way or that way? I don't know. Maybe if I asked this person, he could give information that would be helpful or even tell me the best route." Here the deliberation dialogue gives rise to the usefulness of shifting to information-seeking dialogue.

Another type of information-seeking dialogue is the media interview. A televised interview of a celebrity, for example, may arise out of interest in the personal character or commitments of the interviewee. The skill of the interviewer is to make the respondent feel relaxed and to ask the right sort of questions to prompt her to come forward with the desired sort of information that will be of interest to the viewing audience.

Another very common type of information-seeking dialogue is searching through a computerized database for some specific item of information. For example, the database might be a collection of titles and abstracts of books and articles in an academic field. The searcher can ask for a specific title or for works by a specific author. But she could also ask, more generally, for all titles on a specific subject. Or she could ask for a combination (conjunction) of topics or for any title that includes any of (a disjunction of) topics. The less specific the information sought, the more likely that the data produced will be larger (and more costly). Practical limitations, in such cases, often require keeping the question as specific as possible, in order to prevent a wasteful *embarras de richesses*.

In much everyday practical reasoning from goals to actions, information-seeking dialogue is mixed with (or joined to) deliberation dialogue. Knowledge, or information at any rate, is important as an ingredient that makes deliberation well informed as a basis for carrying out a task.

6. Deliberation

Deliberation is a type of dialogue that arises out of a need to consider taking action. Sometimes the initial situation is a practical problem posed by a question like "How do I do this?" In other cases, the initial situation is posed by a practical conflict, where there are two (or more) opposed actions or ways of doing something, and a choice between them needs to be made.

The main thread of reasoning that holds argumentation together in deliberation is practical reasoning, a kind of goal-directed reasoning that concludes in an imperative to action.[18] One type of premise in a practical inference is the goal premise; the other is the means premise, which is based on the agent's knowledge or information of the particulars of his individual circumstances. The means premise says, "This is the way to carry out the action, given the resources and information available to me." The two premises lead toward a conclusion describing a prudent (practical) course of action for the agent based on the assumptions made in the premises. The goal of deliberation is to reach such a conclusion or decision on how to act prudently in a given situation.

According to Aristotle (1968; *Nicomachean Ethics* 1112 a 30–1112 b 1), "we deliberate about things that are in our power and can be done," and hence "in the case of the exact and self-contained sciences there is no deliberation." Theoretical wisdom is the appropriate kind of wisdom in these sciences, whereas practical wisdom is appropriate for the practical sciences (or arts), like medical treatment and money-making (1112 b 4). According to Aristotle (*Nicomachean Ethics* 1141 b 11), "no one deliberates about things invariable," and practical wisdom is concerned with particulars that are subject to change.

Accordingly, Aristotle is led to the conclusion (1142 a 23) that practical wisdom is not scientific knowledge. Furthermore (1142 b 1), he concludes that there is a difference between inquiry and deliberation. Excellence in deliberation, he thinks (1142 b 15), is a kind of *correctness of thinking*, something that involves reasoning as well as searching for something and calculating.

Deliberation is carried out on the basis of information, but a good deal of that information describes the particular circumstances of the agent's given situation, something that is constantly changing.

Hence deliberation, by its nature, is constantly subject to revision and updating as new information comes in. For this reason the conclusion of a sequence of practical reasoning in deliberation is generally best regarded as a tentative presumption, a defeasible proposition subject to rebuttal if the situation changes. Consequently, it is important in deliberating to be open to new information and not to be dogmatic, or too fixed in one's preconceptions.[19] A certain flexibility is good, and judgment is needed to weigh the value of presumptions in a situation where hard knowledge is lacking. The kinds of skills of excellence of reasoning that are most useful in deliberation are therefore somewhat different from those that are most important in the inquiry. The inquiry involves a very conservative style of thinking that strives for high standards of proof in order to avoid error. Deliberation should not become too conservative or it runs the risk of losing flexibility.

Deliberation is often functionally joined to, and dependent upon, inquiry or information-seeking dialogue, because the second (means) premise of a practical inference is based on knowledge or information. A good example is the way political deliberation is often dependent on scientific knowledge derived from consultation with expert advisers. Or to take another kind of example, a tourist trying to get to the Central Station in a foreign city may have to depend on information acquired by asking directions of a passerby.

It may seem strange at first to think of deliberation as a type of dialogue, for much ordinary deliberation appears to be solitary. Still, even solitary deliberation can often be very well described as a kind of dialogue with oneself, where questions are posed and replied to, where critical doubts are raised, and two sides of a proposal are played off against each other by argumentation pro and contra.

At the other extreme, much deliberation, for example, the kind that takes place in political debating, seems to be a group activity involving more than just two participants. Even so, however, cases of this kind of deliberation can often be reduced to two sides, a pro and contra with respect to some contemplated course of action. Even though there are many participants involved, the dialogue can be examined from the point of view of a deliberation by seeing the argumentation as being directed toward supporting the one side or the other. So in the case of either single-person deliberation or multiple-person deliberation (of more than two participants), we can view the argumentation from a perspective of a deliberation in our sense, meaning a normative model of two-person dialogue, where the two participants represent the two opposed sides on the issue of the right course of action to be taken.

The closing stage of a deliberation dialogue is often dictated by practical constraints. Often there is not enough time to acquire enough knowledge to resolve the question definitively one way or the other, by means of an inquiry. Even so, action may be necessary, for even no action (doing nothing) may be a form of action, in the sense of having significant consequences. For this reason, deliberation typically involves presumptive (opinion-based) reasoning, and closure is based on a weight of preponderance of presumption. The burden of proof in a deliberation should not be unrealistically high. And it may often be necessary to reopen deliberations should a situation change.

Deliberation is based on (known or presumed) facts as well as values. Since goals are also very often subject to change and modification as deliberation proceeds, the matter of an agent's values can also be subject to revision in practical reasoning. This is another reason why practical reasoning is a dynamic kind of argumentation that is defeasible in nature in the context of deliberation.

7. Dialectical Shifts

In passages of discourse in everyday conversation, there is quite often a *dialectical shift*, where during the course of argumentation, there is a change from one type of dialogue to another. For example, in the following case, Karen and Doug are riding along on their bicycles, discussing the issue of whether it is better to live in a condominium or a house.

Case 62

Doug: Yes, in a house there is a lot of yard work to do, but with a condominium you can sometimes hear the neighbors.

Karen: I agree, and condominiums have those large fees for maintenance. Oh! The sign just ahead says that the bicycle path goes this way to Lisse, and that way to Sassenheim. Do you want to go to Lisse or Sassenheim?

In this case, a dialectical shift occurs when Karen says 'Oh!' Before that Karen and Doug were engaging in a critical discussion on the topic of whether houses or condominiums are better. After the shift, they began a deliberation type of dialogue on which village they wanted to go to that day.

The dialectical shift in case 62 could be described as a kind of interruption of the first dialogue posed by the practical need for a decision. Even so, there is nothing inherently fallacious or erroneous about the shift. For once this practical question has been decided, Karen and Doug can then resume their critical discussion of the houses-versus-condominiums issue. There can be a kind of implicit agreement to discontinue the first dialogue temporarily, to have a brief interruption for a different type of dialogue to take place. Here we have a shift, but it is not necessarily a bad shift.

In case 62, however, the two dialogues really have nothing to do with one another. Hence the shift can be described as a kind of interruption. In another type of shift, by contrast, the two dialogues can be functionally related to each other.

In the following case, Maurice and Heather are having a critical discussion on the ethics of euthanasia. Maurice maintains that euthanasia should never be allowed under any circumstances, and Heather opposes that point of view.

Case 63

Maurice: If you allow euthanasia in any form, it could lead to people being killed for political reasons, or by greedy relatives.

Heather: Not if it were purely voluntary. The person who elects to die must clearly be doing it of her own free will and not by reason of pressure or coercion from someone else.

Maurice: But that would never work. It's just not practical, and it would be abused by people who would exploit the system.

Heather: Well, in fact, it does work in the Netherlands. There, patients with a terminal illness can elect voluntary euthanasia, in consultation with their physician. The system works there. People are happy with it, and there have not been worrisome complaints of abuse.

Maurice: Well, how can you prove that?

Heather: I have a report here from a Dutch medical journal, written by a Dutch physician who has a good deal of experience with the euthanasia practices in the Netherlands. And it is clear from what he writes that the system is working there and does not suffer from widespread abuses of the kind you are worried about.

This dialogue began as a critical discussion on the issue of whether euthanasia should be allowed or not as a practice. But then the discussion turned to a subissue of whether a system of euthanasia could be practical or whether it would be abused. To bring evidence to bear on this issue, Heather appeals to a report written by an expert who has direct knowledge of a case in point. Thus there has been a shift here from a critical discussion to a type of information-seeking dialogue that could be called an expert consultation. The expert is not actually engaged in the verbal dialogue, but his article is cited as a reliable source of expert knowledge that is relevant to the issue of the critical discussion.

In this case, the information-seeking dialogue is functionally connected to the critical discussion. By bringing in empirical knowledge through the using of an expert source, Heather has thrown some light on the issue of practicality being discussed with Maurice. This will help improve the critical discussion on euthanasia by making it more informed, bringing it into line with current knowledge and developments.

In this case, the appeal to expert opinion is not fallacious, because Maurice is free to question or dispute the article or the qualifications of the person who wrote it or otherwise to continue his argument against euthanasia. Maurice could even bring in his own sources of expert opinion who disagree, if he wishes. In this case, the expert consultation dialogue is not just an interruption to the critical discussion. It is, like the first case, a temporary shift to a different type of dialogue. But the second dialogue functions to assist the intelligent discussion of the subissue in the first type of dialogue.

In both cases above, the shift was temporary and happened adventitiously during the course of the first dialogue. But in other cases, there may be an agreement or announcement that closes off the one dialogue and initiates the other. For example, a group of business people may be having a meeting on whether or not to diversify into a new line of farm implements. At the end of the meeting, the chairman may declare the meeting over and call everyone to adjourn to the bar, where they all begin to discuss recent developments in the Soviet Union. In this type of case, there has been a definite shift, but there is no functional relationship between the two dialogues. Moreover, it is not an interruptive type of shift, because the first dialogue has been (properly) closed off and is not meant to be continued after the session in the bar.

8. Illicit Dialectical Shifts

Dialectical shifts are not always problematic or a sign of an error or fallacy. But they do become a problem, from a point of view of the critical analysis of argumentation, where there is deception or misunderstanding involved. This can occur when one party to the dialogue is unaware of the shift and the other party is trying to conceal the shift or take advantage of the first party's confusion.

A case of this sort concerned a type of television program called an *Infomercial*, which has the format and appearance of a talk show but turns out to be a half-hour commercial. Prior to 1984, the U.S. government had set limits on the length of a commercial. But when these limits were removed, it became profitable for television stations to fill in blank slots late at night with infomercials rather than a movie, for which they would have had to pay.

The tricky thing about infomercials is that they exploit the viewer's initial expectation that he is watching a news or talk show that is presenting information in a reporting or interviewing format. Not until the viewer watches the program for a while does it become clear that the program is really an advertisement for a product. According to a *20/20* report (1990, 13), infomercials are designed to create this deception by appealing to viewers' normal expectations from watching news programs and talk shows on regular programing in the past: "Yes, they have all the trappings, like 'expert' panelists and breaks for commercials, even closing credits, but in truth, they are just commercials, half-hour commercials." The way these programs are made indicates they are exploiting a standard format for one type of news reporting or information presenting type of dialogue to try to deceive the viewer into watching a lengthy sales pitch. The sales pitch is really a different type of dialogue altogether, a sort of one-sided promotion to persuade a viewer to buy something. This type of dialogue is not supposed to be unbiased, or to present both sides of an issue, as news reporting is supposed to be. Hence the shift in this type of case is concealed and involves deceit.

John Stossel, the interviewer, reports on the case of an Infomercial for a "cooking stone," a piece of rock that stays hot after being heated in an oven, so that you can cook on it later. The attractive people who praise the cooking stone in the program are, in reality, all actors from a local talent agency. In the program, they pose as neighbors of the chef who demonstrates the product. Stossel (*20/20*, 14) points out where the deceit lies in this type of program.

> There's nothing inherently wrong with selling through a half-hour commercial unless there's deceit involved. And that's the problem. As we've watched the infomercial business grow, deception's one thing we're seeing lots of. It comes in two forms. First, some infomercials push products that don't do what they say they'll do. And second, the format itself can be deceptive. When you make a commercial look like a talk show, aren't you trying to fool people to make them think that these kinds of endorsements are spontaneous? The man on the right, Mike Levy [sp?], who appears to be just another talk show host, is actually president of the company that produces what's probably the most recognizable infomercial series, "Amazing Discoveries." He sells exciting products, like this unbelievably powerful mixer and a product that will protect you so well, you could set it on fire.

The key thing that accounts for the deception is the shift from one type of dialogue to another. There is nothing wrong per se with a sales pitch, a commercial advertisement for a product. But if the producers are trying to disguise the sales pitch by putting it in another format, this is quite a different matter. The argumentation in the sales pitch is not fallacious or open to critical condemnation per se, just because it is a sales pitch. We all know and expect that a sales pitch is taking a one-sided approach of promoting a product, making no pretense of being unbiased reporting of the assets as well as the defects or shortcomings of the product (in the way we would expect, for example, of *Consumer Reports*).

But if the program is supposed to be a news report and presents itself as such, then that is a different thing. According to the *20/20* report (15), one infomercial presenter even introduced himself as "your Inside Information investigator." According to Stossel (*20/20*, 16), this program, "Rediscover Nature's Formula for Youth," was deceptive because it used terms like "investigative team" to suggest that it was a regular news program. When confronted with the allegation that he was "pretending" to be "a news program to sell a product," the producer replied: "It's called advertising. It's called propaganda. That's the name of the game. Come on, John, it's the real world" (16). This reply attempted to attack Stossel by saying that the news programs he is involved in are also paid for by commercials.

The key difference here to be emphasized is that in regular news programs, the commercials are kept separate from the news program itself. The viewers know what to expect when they are watching a commercial as opposed to a news report. Or at least, the format clearly enables them to be aware of this difference in the type of dialogue that the presenters are supposed to be engaged in. With the infomercial, there is a deceptive shift from the one type of dialogue to another, within the very same program, the very same sequence of argumentation.

A case like this can be called an illicit dialectical shift. The problem lies not in the argumentation itself per se. There is nothing wrong with a sales pitch, necessarily, just because it is a sales pitch. But if that argumentation occurs within a context of dialogue that is supposed to be an objective news report, a presenting of information, and even encourages the viewers to take it this way, then there is something wrong. It is a calculated deception—a dialectical shift that makes the argumentation subject to critical condemnation.

The criticism in such a case pertains not just to gaps or errors in the reasoning in the argumentation. It is a question of the context of dialogue in which that argumentation was put forward. The critical evaluation should take place by looking back to the original type of dialogue from which the shift took place. We need to ask what the original type of dialogue was that the participants were supposed to be engaged in and evaluate the argumentation from this standard. Advertising may be perfectly reasonable if the dialogue is supposed to be a sales pitch. But if it was really supposed to be a critical discussion, or a presenting of information as "news," then it should be evaluated by standards appropriate for that type of dialogue. From that point of view, it may fall far short of standards of good or correct argumentation and may be open to critical questioning and objections. And if the shift is concealed, intentionally or otherwise, that can be a serious problem for critical analysis of the argumentation.

In this case, the shift is illicit because the viewers' expectations that the program watched is engaged in a certain type of dialogue are being deceptively exploited. The advertisers who make up infomercials would argue that it is not an illicit dialectical shift, no doubt. But their arguments are implausible, because it is clear that one major factor in making for the effectiveness of infomercials is that the viewers (or at any rate the less sophisticated ones) see the program as some sort of news report. The producers of these programs do not announce, at the beginning of the program, that what follows is a commercial ad for a particular product. Evidently, the reason they do not do so is that they feel it would lessen the impact of their argument.

Hence this case is a good example of a dialectical shift--a fairly obvious and clear instance of one, once we see what is going on. Shifts associated with fallacies are typically more subtle and covert.

9. Double Deceptions

One of the most problematic types of shift cases is one where both participants wrongly assume that the other party is engaged in a particular type of dialogue. It is a dual misunderstanding. For example, one party may think that the other is engaged in a critical discussion, while the other thinks the first party is engaged in a quarrel.

The quarrel is associated with bias and dogmatism, with an emotional attachment to one's point of view and a tendency to see the issue in absolutes of "us" against "them." The quarrel is often associated with fallacies like "hasty generalization," "black-and-white thinking," and "special pleading" and with bias. Such an association is easy to understand once we realize that the quarrel, as a type of dialogue, is characterized by a rigidity of attitude. But in some cases, this attitude may not be evident to one participant.

A religious zealot, or cult adherent, may appear to be engaging in a critical discussion with a potential convert, for example. The potential convert may think he is engaging in a critical discussion on religion. But in fact, the cult follower may not be open at all to conceding any argument that might give evidence that his religious point of view is wrong or false. He is never really open to defeat at all, and hence his surface appearance of engaging in some sort of persuasion dialogue is really a pretense. This type of failure of communication can be very serious.

In other cases, eristic dialogue is associated not with some relatively constant or permanent group dogma or bias but with some underlying strong feelings of loyalty that surface temporarily in an emotional moment. In such cases, a participant in dialogue may normally be a very careful and critical reasoner who is openly looking at both sides of an issue in a sensitive and thoughtful way. But some particular topic in a given situation may trigger strong emotional feelings that give rise to a quarrel on one side of a dialogue.

As noted in section 1 above, the critical discussion as a type of dialogue requires a willingness to subject one's opinions to critical doubt and an openness to conceding refutation if one's point of view is confronted with a reasonable argument that goes against it. Sometimes, however, due to the frame of mind of one participant, the requirements for this type of dialogue are not present. Sometimes one party tries to engage in a critical discussion, but the other party is so biased, or so strongly caught up in his own point of view, that he cannot even consider the thought of changing it or admitting even the most reasonable qualifications to it, much less abandoning it. In such cases, the one side may have a critical discussion in mind, while the other side engages in eristic dialogue.

In the following case, General H. Norman Schwarzkopf had been transferred back to Washington in 1970 and had invited his sister Sally over to dinner. After cocktails and a long dinner, during which they drank a magnum of champagne, they set down to watch a Korean war movie where several soldiers were caught in a minefield. Schwarzkopf reacted very emotionally to the movie because, as a soldier in Vietnam, he had been caught in a minefield and wounded by shrapnel while trying to rescue another man. He started to say to the soldiers in the movie, "Don't do that," when Sally asked if he was overreacting, saying that the Vietnam War is now behind us. The dialogue following this remark is quoted below from Schwarzkopf's autobiography (1992, 214–15).[20]

Case 64

Sally was looking at me in amusement. "Come on, Norman, it's just a movie. It's not even about Vietnam. Aren't you overreacting?"

"I'm not," I said. I was shaking.

"Why worry about it? It's behind us."

I deeply resented that. "It's *not* behind us. It's still going on. Goddammit, I can't stand the people in this country who say it's over, who are trying to put it behind us, who are trying to pretend it never happened! Don't tell me I shouldn't react. You sound just like the peaceniks!"

Sally misread how strongly I was reacting. She thought I was just being argumentative and pressed on: "You can't just dismiss everything the peaceniks say. They have some legitimate points."

I couldn't believe my ears. I'd always thought Sally was on my side. But what I was hearing was a dismissal of the war and a willingness to walk away from everything we stood for in Vietnam—an attitude that, to my mind, was contributing to the loss of more American lives. I couldn't tolerate that. "I'm sorry," I interrupted, "but if you honestly believe these things, if you honestly feel that way, then I don't want you in this house."

Sally bristled. "Well, I honestly do feel that way."

"Then get out." I was in tears because I felt so betrayed, and now she was crying, too. "Get out of my house."

"Oh, now, Norman, I . . . "

"There's nothing to talk about! Get out."

From the perspective of 1970, Sally's contention that the "peaceniks" had some legitimate points would seem to be an easy concession to make. She was not saying that those who were for pulling out of Vietnam were right absolutely but only that they had "some legitimate points."

Schwarzkopf however, was in an emotional frame of mind and, having just been reminded of his service in Vietnam, reacted emotionally, framing the issue as "my side" against the "peacenik" side he saw as responsible for loss of American lives in Vietnam. He reacted with a quarrelsome burst, feeling betrayed, and tried to stop any further discussion.

The next day, Schwarzkopf felt he had treated his sister badly, and resolved not to let alcohol take control and adversely affect his family relationships. In retrospect, he felt he had reacted inappropriately. But it is not hard to see how, carried away by strong emotional feelings of the moment, and identifying with a cause or point of view that involves strong emotions and loyalties, it is quite common for people to react in an "us against them" polarized and quarrelsome way when confronted with arguments that oppose a deeply held point of view.

In this kind of case, there was no deliberate attempt at deception on one side, and it would be pointless to try to fix blame on one side or the other for what happened. It was a dual deception where both sides wrongly assumed that the other side was engaging in the same type of dialogue the first side was engaging in. This kind of failure to communicate is inevitably futile, because the argumentation of the one side is not even really interacting at all with the argumentation of the other side. Hence nothing can be resolved, and in this case, the exchange ended in tears on both sides.

Schwarzkopf writes that his sister "misread" his reactions and thought he was "just being argumentative." She assumed he was engaging in a critical discussion on the politics of the Vietnam War. On the other side, he thought what he "was hearing" was a quarrelsome attack on himself and his deepest personal values. He thought she sounded like "the peaceniks" and was no longer on his "side." He saw this as a "dismissal" and a "walking away from" everything that his side "stood for in Vietnam." He saw her argumentation as a personal attack and betrayal and lashed out in what he thought was a continuation of this eristic dialogue, asking his own sister to "get out," a very harsh thing to say. At the end, with both in tears, he shut off the dialogue, saying, "There's nothing to talk about!"

We see in this kind of case the problem that a critical discussion is not always appropriate in a given situation where, for example, one party may have very strong feelings on an issue without the other party's realizing how the first party is taking the conversation. This type of case can be the most serious type of illicit shift because of the deception on both sides and the confusion engendered by the illusion, on both sides, that the speakers are interacting together in some sort of really connected sequence of dialogue exchanges.

10. Mixed Dialogues

Some other familiar types of dialogue can be classified as *mixed dialogues*, or cases where two different types of dialogue are mixed together in the same case. Sometimes these cases of mixed dialogues have to be approached carefully, because special circumstances affect the normative rules that need to be taken into account in judging the argumentation in a given case.

A good example is the type of dialogue called the *debate*, where two opposed sides are argued out on an issue, and the winning side is judged by some third party—a referee, moderator, or audience. The debate appears to be a critical discussion when we first look at it. But the problem is that debaters can score good points and can win over a judge or audience successfully even while using bad or fallacious arguments that would violate the rules of a critical discussion. It is clear, then, that a debate is not exactly the same type of dialogue as a critical discussion.

The debate has a strong adversarial aspect. The idea is to let the debaters fight it out in a free arena in which both can bring forward their most powerful arguments and then see who wins. In this respect, the debate is best classified as being an eristic type of dialogue—at any rate, more so than the critical discussion is allowed to be. Thus we could say that the debate is a mixture of the critical discussion and the quarrel.

At its best, the debate can be a noble thing that has the good qualities of a critical discussion, by bringing out the real positions and the most convincing arguments on both sides of a controversial issue. In other cases, however, the quality of discussion in a debate in this regard can be very poor. In such cases, for example, ad hominem arguments that would be highly fallacious in a critical discussion are very successful arguments in a debate, scoring heavily with the audience who is to judge the outcome.

Another factor about debates, like some other cases of argumentation, is that they need to be evaluated in relation to the social or institutional setting in which they take place. A particular debate may have certain rules laid down in advance that determine who will speak when, what they can say, and how the outcome will be judged. In a forensic debate, these rules may be laid down at the outset and the participants may agree to them before the start of the argumentation stage. In a political debate, the rules may be set down in a handbook of parliamentary rules, for example, and these rules may be enforced by an appointed speaker of the house.

A *speech event* is a particular social, cultural, or institutional setting having rules and expectations for the conduct of argumentation that the participants are bound to follow by taking part in dialogue in this setting. For example, if the argumentation is taking place in a parliamentary debate, then the participants are bound to follow the rules adjudicated and enforced by the Speaker of the House. These rules are generally codified in handbooks that the participants can consult. A good example would be the question period of the Canadian House of Commons debates.

Normally, the House of Commons is nearly deserted during parliamentary debates, but during question period, which takes place five times a week when the House is in session, the room is crowded. Possibly the reason is that most of the television and other news media coverage comes from the exchanges that take place during question period. According to Franks (1985, 3), there is a remarkable exodus after question period: "Where there were 280 members there are now twenty-five; where the press gallery was packed, only two or three remain; the public galleries are empty." Regular debates tend to be dull, whereas in question period, aggressive, often personal attacks on government ministers by the opposition members are often good theater, played over many times on television news reports.

The purpose of the question period is to allow the opposition members to ask the government ministers for information and to allow the opposition members to press for action on matters that are of urgent concern. The type of dialogue is mixed. It is supposed to be an information-seeking type of dialogue and an action-directed type of dialogue.

In fact, however, *Beauchesne* (1978), the book of parliamentary rules, lists many restrictions, including the following (131): questions may not ask for a legal opinion, may not inquire into the correctness of a statement made in a newspaper, may not require too lengthy an answer, and may not raise a matter of policy too large to be dealt with as an answer to a question. Questions should also be brief, should not be based on a hypothesis, and should not "cast aspersions" on anyone (132). The interpretation of these rules is, in practice, quite permissive. Lengthy questions, questions based on hypotheses, and questions that attack the character of the respondent are quite often tolerated.

Government respondents have time to do research to answer written questions. But they must answer oral questions on the spot. The Canadian House of Commons appears to be unique among parliaments in Western democratic countries in requiring impromptu answers in the oral question period. The result is that the question-reply exchanges are often lively and argumentative. A government minister can decline to answer a question according to *Beauchesne* (1978, 133), but ministers rarely do. Usually a reply that at least addresses the question is given.

A question or reply that violates the rules of parliamentary procedure will be ruled out by the Speaker of the House. The Speaker is a kind of moderator, elected by a majority of the members of the House. If a member persists in breaking a rule, he or she may be asked by the Speaker to leave the House. In some instances, the Speaker may ask a member to apologize for unparliamentary behavior.

Political debate is typically a complex mixture of all six types of dialogue. It is highly eristic and partisan, often concealing aspects of interest-based negotiation. Often, as well, it concerns deliberation on what should be done, on the basis of expert consultations, other types of information-seeking dialogue, or inquiries. Even so, it is a presumption in democratic countries that political debate should at least be bound by some requirements of a critical discussion. The presumption is that the more outrageous fallacies that are lapses of critical discussion should be criticized and not tolerated.

In criticizing arguments in political debates on the grounds that fallacies have been committed, we can look at such a debate from the point of view of a critical discussion. This viewpoint entails a conditional analysis of the argumentation, meaning that the analysis postulates that *if* the discourse is evaluated as if it were supposed to be a critical discussion, such-and-such fallacies in it can be criticized as shortcomings.

Because participants in argumentation are often unclear as to what type of dialogue they are supposed to be engaged in, criticisms of fallaciousness must often be conditional in nature. Even so, however, such criticisms can have force, because arguers may be quite effectively criticized if it is pointed out that their argumentation does not meet the requirements of a critical discussion. Then it is up to them to say whether they think that this is the type of dialogue that they are supposedly taking part in. Then the charge of fallacy turns on the dialectical question of what type of dialogue the arguers should be engaged in.

5

Argumentation Schemes

Twenty-five argumentation schemes are presented in this chapter of a kind that are (1) presumptive in nature, (2) all very common in everyday conversation, and (3) all related to one or more of the major informal fallacies. This list is not meant, in any sense, to be complete. Kienpointner (1992) lists many more argumentation schemes and includes a goodly number of those treated here. The intent of this chapter is to present these schemes in a concise way that will be useful in analyzing fallacies and in understanding the structure of the concept of fallacy generally.

A pioneering account of argumentation schemes (called "modes of reasoning") was given by Hastings (1963). Many other argumentation schemes can be found in Perelman and Olbrechts-Tyteca (1969). They distinguished between "associative" or positive argumentation schemes, used in support of one's own argumentation, and "dissociative" or negative argumentation schemes, used to attack another party's argumentation. Kienpointner calls the latter type *Gegensatz* schemes, and I often call them refutation schemes below.

Certain of the argumentation schemes treated in this chapter (some are noted specifically under this heading, but others also bear important relationships) could be described as subschemes of a broad and very common kind of reasoning called practical reasoning in the sense of Clarke (1985), Audi (1989), and Walton (1990a). Practical reasoning, as opposed to purely theoretical or discursive reasoning, is used to reason toward a practically reasonable or prudent course of action on the basis of one's goals and knowledge of the circumstances of one's given situation.

Practical reasoning is a chaining together of two basic schemes of practical inference, called the necessary condition scheme and the sufficient condition scheme (respectively, below). *G* is a goal, *a* is an individual agent, and *A* is a state of affairs.

> *G* is a goal for *a*.
> Bringing about *A* is necessary (sufficient) for *a* to bring about *G* (as far as *a* knows).
> Therefore bringing about *A* is prudentially right as a course of action for *a* to take.

When used in dialogue, practical reasoning results in a sequence of argumentation chaining together necessary and sufficient inference schemes of the kind shown above.[1] As such, practical reasoning is often an overarching structure or "master scheme" into which other argumentation schemes fit.

Matching either of the argumentation schemes above are four key critical questions.

1. Are there alternative means of realizing *G* other than *A*?
2. Is it possible for *a* to bring about *A*?
3. Does *a* have goals other than *G*, goals that may even be incompatible with *G*?
4. Are there negative side-effects (consequences) of *a*'s carrying out *G* that should be taken into account?

In addition to these four critical questions, there are also critical questions matching each premise.

5. Is *G* really a goal that *a* is committed to?
6. Is bringing about *A* necessary (sufficient) for *a* to bring about *G*?

Practical reasoning functions in a dialogue to alter a participant's commitments. It is very commonly used in deliberation, on deciding on a prudent course of action for an agent in a given situation. But it is also used in other types of dialogue, like information-seeking dialogue and critical discussion.

Practical reasoning is inherently presumptive in nature, because an agent's knowledge of its situation tends to be incomplete and based on rapidly changing, imperfectly known information. Goals can also change and are often difficult to determine, except by conjecture. In practice, therefore, the schemes for practical reasoning shift a burden of proof or disproof to one side or the other in a dialogue where opinions are divided on how best to proceed. Often instead of going for best or "maximizing" solutions, practical reasoning can be satisfied with outcomes that do well enough to get the job done (so-called satisficing solutions).[2]

Practical reasoning as a type of logical structure had been pretty well ignored by scientists and formal logicians (and still is, by many of them) until the recent advent of research on robotics and artificial intelligence.

1. Presumptive Reasoning

Deductive and inductive reasoning is to be distinguished from presumptive reasoning by the nature of the link between the premises and the conclusion, as used in an argument, and by the nature of the warrant, or linking (general) premise that connects the premises to the conclusion. In a deductively valid argument, if the premises are true, then the conclusion must be true, in every case. In an inductively strong argument, if the premises are (probably) true, then the conclusion can be evaluated as likely to be true, with a certain degree of probability. Both of these types of arguments can be judged for validity (or conditional probability in the case of inductive arguments) by means of a calculus that can be applied to the argument independently of the context of dialogue surrounding it.

Presumptive reasoning is evaluated, in contrast, by its use in a context of dialogue where two parties are reasoning with each other. A presumptive argument is judged by whether it shifts a weight of presumption to the side of the other party in a dialogue. Presumptive reasoning is always tentative or provisional in nature.[3] In presumptive reasoning, an argument advanced by a proponent shifts a weight of presumption by fulfilling the requirements for the use of that argumentation scheme in a context of dialogue, placing an obligation on the respondent to reply by raising critical doubts appropriate for that argumentation scheme. Presumptive reasoning is inherently defeasible in nature, meaning that it is suppositional and is subject to defeat by exceptional cases.

Normally when an assertion is made by one party in a critical discussion, the proponent becomes committed to the proposition asserted, in a strong sense of 'commitment' implying a burden of proof. By contrast, a party is free to make a (pure) assumption (supposition) without incurring a burden of proof to back it up by evidence if challenged. Presumption, according to the analysis given in Walton (1992c, chap. 2; 1992a) is a speech act halfway between assertion and assumption. A presumption is put forward "for the sake of argument," for practical purposes to allow a dialogue to go forward on a provisional basis, where there is not enough evidence to prove conclusively the proposition presumed or to disprove it.

According to the analysis given in Walton (1992c; 1992b) the speech act of presumption reverses the roles of the proponent and the respondent in a dialogue. When a proponent puts forward a presumption in argumentation in a dialogue, she has no burden to prove it, in order to maintain it as a presumption. Instead, the respondent has a burden to disprove it, if he wants it to be dropped as a presumption in the dialogue. But if the respondent does come forward with evidence that is sufficient to refute the proposition contained in the presumption, then the proponent is obliged to retract that proposition as a commitment in the dialogue.

Presumption is a very useful device in argumentation because it enables a dialogue to move forward even where there is insufficient evidence available at a given point in the dialogue to prove a proposition or to assert it categorically. It allows you to be able to make tentative concessions to your opponent, for the sake of argument, to see where the argument might lead. Some presumptions are made in practical reasoning and deliberation, on grounds of safety, for example, to allow a prudent decision or line of action, where opinions are divided on the best way to proceed.

Case 65

Vince and Adele are collecting mushrooms in the woods to put on their pizza for dinner. They are not botanists or any sort of experts, but they are familiar with the kind of mushroom they usually collect. Vince picks up one "mushroom" that looks a little different. Given the remote possibility that it is poisonous, Adele proposes not including it with the mushrooms for pizza.

Here safety suggests acting on the presumption that this "mushroom" is or could be poisonous and tossing it aside.

Presumptions are common in legal reasoning. For example, for purposes of distributing an estate, an individual may be presumed dead if there is no evidence that he has been alive for a fixed period, for example, seven years. To rebut such a presumption of death, the party who contends otherwise must produce some evidence that the individual is still alive.

Presumptions are often agreed to by two parties in order to minimize the need for subsequent dialogue.

Case 66

If I don't hear from you by Friday, I will assume that you will be coming
to the reception for the Dean on Saturday.

Presumptive argumentation is subject to agreement by both parties
in a dialogue and can be canceled by either one.

Presumptive reasoning is based on normal expectations and is subject to defeat in exceptional cases. Thus as a kind of reasoning, it is
not highly reliable in many cases. One must always be open to giving
it up, if new information comes in, showing an attitude of flexibility
and sensitivity to qualifications.

All twenty-five argumentation schemes in this chapter are inherently presumptive as types of argumentation. Fulfilling the requirements of an argumentation scheme in a context of dialogue draws a
weight of presumption to the proponent's side of the dialogue. Each
argumentation scheme has a set of matching critical questions, however. Asking one of these critical questions removes that weight of
presumption, at least temporarily, until the proponent provides an
adequate answer. In any given dialogue, there is a global burden of
proof on both sides, set at the initial stages of the dialogue. The use
of argumentation schemes and matching critical questions affects
the fulfillment of these burdens by distributing local weights of presumption. Thus the context of dialogue is crucial to evaluating argumentation schemes as they are used in a given case.

2. Case-Based Reasoning

In argument from example, a particular case is cited in support of
a presumptive generalization of the form: if an individual x has property F, then x typically or normally (subject to exceptions) also has
property G. As an example, suppose someone is arguing that tipping
leads to misunderstandings and embarrassment and cites the following example to support her contention.

Case 67

Well, one time my husband failed to leave a tip for our coats to be
checked, before the meal, and the waiter spilled soup on his suit. My
husband was angry because he thought the waiter did it on purpose
because he failed to tip. But it was unclear whether you were supposed
to leave the coat-checking tip before the meal or afterward.

Argument from example shifts a weight of presumption in favor of a
conclusion but is subject to critical questioning and to argument us-

ing a counterexample. The argumentation scheme for argument from example is the following.

> In this particular case, the individual a has property F and also property G.
> Therefore, generally, if x has property F then x also has property G.

There are five critical questions corresponding to this argumentation scheme.

1. Is the proposition claimed in the premise in fact true?
2. Does the example cited support the generalization it is supposed to be an instance of?
3. Is the example typical of the kinds of cases the generalization covers?
4. How strong is the generalization?
5. Do special circumstances of the example impair its generalizability?

Argument from example is inherently subject to qualifications with respect to the individual features of a given case and is therefore inherently susceptible to the secundum quid fallacy of neglect of qualifications.

Argument from analogy is used to argue that a proposition is true in a given case on the grounds that it is true in a similar case. The argumentation scheme for argument from analogy is the following.

> Generally, case C_1 is similar to case C_2
> A is true (false) in case C_1.
> Therefore A is true (false) in case C_2.

Similarity is always similarity in certain respects. But, of course, any two (different) cases will also be dissimilar in certain respects. Therefore, argument from analogy is always inherently presumptive in nature, subject to rebuttal by the citing of some new circumstances of a case.

The following case occurred in the context of an article relating how the author's mother stayed home from work to look after her as a child (Chazin 1989, 32). The concluding part of the article used an analogy.

Case 68

A few months ago, my mother came to visit. I took off a day from work and treated her to lunch. The restaurant bustled with noontime activity as business people made deals and glanced at their watches. In the middle of this activity sat my mother, now retired, and me. I could see from her face that she relished the pace of the work world and I wondered how this same woman had managed to spend years as a full-time mother.

"Mom, you must have been terribly bored staying at home when I was a child," I finally said.

"Bored? Housework is boring. But you were never boring."

I didn't believe her, so I pressed. "Surely children are not as stimulating as a career."

"A career is stimulating," she said. "I'm glad I had one. But a career is like an open balloon. It remains inflated only as long as you keep pumping. A child is a seed. You water it. You care for it the best you can. And then it grows all by itself into a beautiful flower. A flower is never boring, no matter how it chooses to grow."

Just then, looking at her, I could picture us sitting at her kitchen table once again, and I understood why, under her love and guidance, it had been so easy to appreciate the dandelions in life.

The issue is whether it is best for a mother to stay home to look after the children or have a career. The argument from analogy draws a weight of presumption in favor of one side.

The critical questions for the argument from analogy are the following.

1. Is A true (false) in C_1?
2. Are C_1 and C_2 similar, in the respects cited?
3. Are there important differences (dissimilarities) between C_1 and C_2?
4. Is there some other case C_3 that is also similar to C_1 except that A is false (true) in C_3?

Analogies are inherently misleading, because no two cases are exactly alike. But argument from analogy should not be classified as essentially fallacious—it can rightly be used to shift a burden of proof in some cases.

Examples and analogies are often used in illustrations, explanations, and other speech acts that are not arguments. One has to be careful to see that not all analogies and examples involve the argumentation schemes for argument from analogy and argument from example.

In argumentation from sign, a particular finding in a given case is taken as indicative evidence that a proposition is true (false) in that case. The sign is a general indicator that is normally or generally linked to a particular outcome or condition and functions as a clue, or basis for a guess that this condition is present. A common type of case is the following.

Case 69

>Bob is covered with red spots.
>Therefore Bob has the measles.

This kind of argumentation depends very much on the circumstances of a given case, subject to collecting more evidence of the case that may confirm or refute the suggestion posed by a sign. For example, if we find out that Bob has just been rolling in poison ivy and has had the measles before, then the conclusion that he has the measles may be dropped.

Argument from sign can be seen as a kind of converse causal argumentation. Presumably, in case 69, the condition of having the disease called the measles is thought to be causing the red spots. This presumed causal link is the basis for drawing the conclusion based on argument from sign. Thus in some cases, argumentation from sign is based on or related to the type of argumentation called argument from effect to cause in section 4 of this chapter.

One of the most famous examples of argument from sign is the classic example of presumptive reasoning in the Indian tradition (Hamblin 1970, 178–80), which could be illustrated by the inference below.

Case 70

>If there is smoke coming from an area, then generally (normally), there is fire in that area.
>There is smoke coming from the hill.
>Therefore there is fire on the hill.

This example has often been cited as the classic case of a presumptive inference, because in some cases, there can be smoke without fire. Even so, on a practical basis of presumption, such an inference could constitute a good reason for calling the fire department.

3. Verbal Classification

Argumentation from verbal classification concludes that a particular instance has a certain property, on the grounds that a verbal classification of the instance generally has such a property. Verbal classifications of particular instances tend to be vague, and so the argument from a verbal classification tends to be defeasible.

An example is the following argument.

Case 71

> Government bonds earn a 5 percent annual interest rate this year.
> Five percent can be classified as a mediocre return.
> Therefore government bonds earn a mediocre return this year.

Hastings (1963, 36) calls this type of argumentation "argument from criteria to a verbal classification."

The argumentation scheme for the argument from verbal classification is the following.

> *a* has property *F.*
> For *x* generally, if *x* has property *F,* then *x* can be classified as having property *G.*
> Therefore *a* has property *G.*

The critical questions matching this argumentation scheme are the following.

1. Does *a* definitely have *F*?
2. How strong is the verbal classification expressed in the second premise?

The generalization in the second premise is presumptive and defeasible, not unrestrictedly universal, because verbal classifications are subject to failure in nonstandard cases. For example, in case 71, the definition of a "mediocre" return can vary quite a bit, depending on the type of investment and the circumstances of the case. Especially the vagueness of classifications of common empirical terms, like 'rich' or 'bald,' brings in borderline cases and exceptions when using this kind of argumentation.[4]

Corresponding to argument from verbal classification, there are two refutational or negative argumentation schemes.

Argument from vagueness of a verbal criterion claims that a verbal classification is overly vague and therefore cannot sustain the conclusion it was supposed to support. In response to case 71, an opponent might use argument from vagueness of a verbal criterion as follows.

Case 72

> The concept of a mediocre return is too vague. What is a mediocre return on one type of investment is not a mediocre return on another. Therefore you can't say that government bonds earn a poor return this year just because they yield a return of 5 percent.

This kind of argument has sometimes been called *the argument of the beard*, especially when combined with gradualistic reasoning. According to Moore, McCann, and McCann (1985, 315), the argument of the beard is using "middle ground, or the fact of continuous and gradual shading between two extremes, to raise doubt about the existence of real differences between such opposites as strong and weak, good and bad, and black and white."

The argumentation scheme for the argument from vagueness of a verbal criterion is the following.

> Some property F is used to classify an individual a in a way that is too vague to meet the level of precision required to support such a classification.
> Therefore the classification of a as an F should be rejected.

This argumentation scheme is refutational (dissociative) because it is used to attack a classification used as the basis of an opposed argument.

This type of argumentation is common in ethical disputes. For example, one party might argue that abortion is wrong because the fetus is a person, using argument from verbal classification. The other party might then reply by saying that you can't define the fetus as a person, because the concept of a person is too vague.[5]

Another refutational scheme is the argument from arbitrariness of a verbal criterion. For example, suppose Bob clarifies his argument that the fetus is a person by making it more specific and Helen attacks Bob's argument as follows.

Case 73

> **Bob:** The fetus is a person during the third trimester.
> **Helen:** You mean that just before the third trimester it is not a person. And then the first day of the third trimester, all of a sudden it is a person. That is arbitrary.

Here Bob has given a less vague criterion, but Helen takes advantage of this to argue that his way of classifying a fetus as a person is arbitrary.

Arguments from or against a verbal classification depend on a term's already having an accepted meaning in common knowledge or linguistic practice. Various other kinds of verbal argumentation are definitional in nature—that is, they proceed by defining a term, often in a way that may depart from existing usage.

Generally participants in a critical discussion will try to phrase their arguments in terms that sound positive for their side and negative for the opposing side. For example, in abortion disputes, both sides describe their position as *pro*, that is, prochoice or prolife. This tendency becomes more pronounced in a quarrel. For example, in wars or territorial disputes, the people on the opposing side are routinely called "terrorists," whereas the people on one's own side are classified as "freedom fighters." The same words (turned around by either side) are used. Thus argumentation from verbal classification should be subject to critical questioning by both sides in a critical discussion.

4. Causal Reasoning

The argument from cause to effect concludes that a particular event or state of affairs will or might occur on the grounds that another event will or might occur that will or might cause it. The future is always uncertain, when dealing with particular cases, so the warrant of argument from cause to effect is always provisional in nature.

According to van Eemeren and Grootendorst (1992, 97), in this type of argumentation, which they call causal or instrumental, "the acceptability of the premises is transferred to the acceptability of the conclusion by making it understood that there is a relation of *causality* between the argument and the standpoint [conclusion]." As an example, they cite the following case (97).

Case 74

Tom has been drinking an excessive amount of whiskey.
Drinking too much whiskey leads to a terrible headache.
Therefore Tom must have a terrible headache.

The 'must' here is clearly a presumptive modality. Tom might not have a terrible headache, in this particular instance, for various reasons. He might be asleep, or he may have taken medications that prevented the headache. He might even be dead. But it would be reasonable to presume, in the absence of any indications that any of these things are true in this case, that Tom can be inferred, by implicature, to have a terrible headache. The Gricean term 'implicature' means that the inference is a reasonable presumption suggested by the context of dialogue but is not a required conclusion logically implied (entailed) by the premises.

The argumentation scheme for the argument from cause to effect is the following, where A and B are states of affairs (propositions describing events).

Generally, if A occurs, then B will (might) occur.
In this case, A occurs (might occur).
Therefore in this case, B will (might) occur.

In this kind of reasoning, although it is variable in strength, generally the possibility of intervening causal variables means that it is a weaker prediction using words like 'may' or 'might' rather than 'will' or 'must.'

The critical questions matching the argument from cause to effect are the following.

1. How strong is the causal generalization?
2. Is the evidence cited (if there is any) strong enough to warrant the causal generalization?
3. Are there other causal factors that could interfere with the production of the effect in the given case?

Argument from cause to effect has the form of modus ponens except that the conditional in the major premise is presumptive, instead of being strict: the presumptive conditional is open to qualifications and exceptions as applied to a particular case.

An example would be the following case.

Case 75

If the Soviet Union breaks up, there will be political instability in Eastern Europe for years to come.
The Soviet Union is breaking up.
Therefore there will be political instability in Eastern Europe for years to come.

Argumentation from effect to cause is similar in nature, but the conditional goes from effect to cause.

Case 76

If a murder victim died from apnea (lack of oxygen), it may be presumed that the person was strangled.
Mr. Smith, a murder victim, died from apnea.
Therefore it may be presumed that Mr. Smith was strangled.

The argument from effect to cause has an argumentation scheme and critical questions comparable to argument from cause to effect. But it is a retrodiction as opposed to a prediction that is expressed by the major premise.

In argumentation from correlation to cause, the premise posits a correlation between two states of affairs and the conclusion infers the existence of a causal connection between them. This type of reasoning has already been analyzed in Walton (1989a, 228–34), where causality is described as a field-dependent relation, meaning that variables are held constant, assuming that the situation is stable or normal, in a particular, given case. This idealization makes causal reasoning presumptive in nature.

The argumentation scheme for the argument from correlation to cause is the following structure.

> There is a positive correlation between A and B.
> Therefore A causes B.

The seven critical questions matching this argumentation scheme are given in Walton (1989a, 230).

1. Is there a positive correlation between A and B?
2. Are there a significant number of instances of the positive correlation between A and B?
3. Is there good evidence that the causal relationship goes from A to B and not just from B to A?
4. Can it be ruled out that the correlation between A and B is accounted for by some third factor (a common cause) that causes both A and B?
5. If there are intervening variables, can it be shown that the causal relationship between A and B is indirect (mediated through other causes)?
6. If the correlation fails to hold outside a certain range of causes, then can the limits of this range be clearly indicated?
7. Can it be shown that the increase or change in B is not solely due to the way B is defined, the way entities are classified as belonging to the class of Bs, or changing standards, over time, in the way Bs are defined or classified?

As these critical questions are answered adequately in a given case, the weight of presumption accorded to the argument from correlation to cause is increased.

Scientific and medical investigations are often initially based on argumentation from correlation to cause. This kind of reasoning is often hypothetical in nature, allowing that the correlation may be a coincidence of some sort, until a systematic analysis of the causal mechanism is found. A good example is the following case from Walton (1989a, 231), originally described in the account of de Kruif (1932).

Case 77

In 1925, pernicious anemia was a fatal disease that caused people to die because their bones mysteriously failed to produce red blood cells. By 1926, Dr. George R. Minot had found through clinical experience that feeding large quantities of liver to forty-five of his patients with pernicious anemia was followed by a great increase in red corpuscle count in each one. Moreover, each of these patients started feeling better and, when kept on a diet of liver, survived to continue a healthy life.

Minot's first reaction (de Kruif 1932, 107ff.) was to suspect a coincidence. It wasn't until much later that laboratory studies found that it was the vitamin B_{12} in the liver that went through the patient's blood and made the bone marrow start producing new red blood cells. Thus the initial presumptive argumentation based on the correlation between restoration to health and ingestion of liver turned out to be vindicated as a good causal argument in this case.

Just by itself, a simple argument from correlation to causation should be treated as presumptively weak in the sense of being open to critical doubts. As the appropriate critical questions are answered, such an argument gains presumptive strength. As such, however, it is a presumptive, not a conclusive, type of argumentation. The conclusion only becomes known to be true, as part of an inquiry, when the causal link is established by the standards and methods for the type of inquiry involved.

5. Commitment-Based Reasoning

In the Hamblin (1970; 1971) structure of formal dialogue, commitment is the basic concept. As the participants make moves, for example, ask questions, make assertions, in a dialogue, a record is kept (in a commitment store) of the propositions each participant is committed to, by virtue of having made each prior move. An analysis of commitment in dialogue has been given in Walton and Krabbe (1995), in rigorous persuasion dialogue (RPD) and permissive persuasion dialogue (PPD). In the former type of dialogue, the commitment store is an explicit set of propositions (called light-side commitments) defined exactly by the rigorous rules and regimented moves of the dialogue. In the PPD type of dialogue, the commitment store is partly composed of so-called dark-side commitments, propositions that exist, are not explicitly known by the participants, and must be conjectured on the basis of presumption.

In the argument from commitment, one participant in a dialogue draws a conclusion expressing an alleged commitment of the other party, based on a premise indicating that party's prior commitment to some proposition in the dialogue. For example, suppose that Jack and Jill are arguing whether or not the monarchy is a good thing to preserve, and Jack has argued that the royal family is vitally important to Britain and ought to be maintained in its present form.

Case 78

Jill: Well, I presume from what you say that you think that the Queen should not have to pay income tax.

Jack: Of course not. I mean, she should not have to pay it.

Without going into detail, we can presume that Jill has inferred this proposition as a plausible consequence drawn from previous commitments expressed by Jack in their dialogue on the monarchy. If Jack were to deny the conclusion and take the opinion that the Queen should have to pay income tax, then there would be a weight of presumption on him to square that commitment with his previous commitments as expressed in the prior dialogue.

The argumentation scheme for the argument from commitment is the following.

> Generally, if a participant P is committed to A, then P can also be expected to be committed to B.
> P is committed to A, as shown by the dialogue.
> Therefore P is committed to B.

The critical questions for the argument from commitment are the following.

1. What is the evidence from the dialogue showing that P is committed to A?
2. How strong is the inference from commitment to A to commitment to B?

In this type of argumentation, it is generally possible (especially in PPD) for the respondent to retract or deny commitment to B. But in order to retract commitment to B, the respondent must either retract commitment to A or dispute the link between A and B as commitments that go together.

The circumstantial argument against the person is a questioning or criticizing of an arguer's commitment by citing a presumption of inconsistency in his commitments. Typically, the inconsistency alleged is a pragmatic (practical) inconsistency rather than a purely logical inconsistency, and the allegation often relates to personal actions or past conduct of the arguer criticized. The term 'circumstantial' is appropriate because the alleged inconsistency is between his personal circumstances and what he says in his argument. Hence the expression "You don't practice what you preach" characteristically expresses the thrust of this type of criticism of this dissociative type of argumentation.

The argumentation scheme for the circumstantial argument against the person is the following.

> a has advanced the contention that everyone in a certain reference class C ought to support proposition A and be committed to A.
> a is in the reference class C.

It is indicated by *a*'s own personal circumstances that he is not committed to *A* (or even worse, is committed to the opposite of *A*).

Therefore *a*'s commitment to *A* is open to doubt.

The upshot of this type of argumentation is the allegation that *a* is himself inconsistent in his own commitments, and therefore *a* must be either confused or insincere. In either event, *a*'s argument is not based on a coherent commitment set, and his integrity (sincerity, honesty) as a participant in the dialogue is open to question.

The critical questions matching the circumstantial argument against the person are the following.

1. What are the propositions alleged to be practically inconsistent, and are they practically inconsistent?

2. If the identified propositions are not practically (pragmatically) inconsistent, as things stand, are there at least some grounds for a claim of practical inconsistency that can be evaluated from the textual evidence of the discourse?

3. Even if there is not an explicit practical inconsistency, what is the connection between the pair of propositions alleged to be inconsistent?

4. If there is a practical inconsistency that can be identified as the focus of the attack, how serious a flaw is it? Could the apparent conflict be resolved or explained without destroying the consistency of the commitment in the dialogue?

The basis of the circumstantial argument against the person lies in the rules for cooperativeness of a conversation of Grice (1975). A participant in a dialogue of the cooperative type (like a critical discussion) is supposed to be sincere in making contributions to the conversation. Inconsistency of commitments is a sign of insincerity on the part of a participant.

The following example of a circumstantial argument against the person is taken from an article (McAuliffe 1980, 51–58; reprinted in Walton 1985a, 267–74). In the segment quoted below (Walton 1985a, 273–74), a representative of the news media, the editor of the *Hamilton Spectator*, is accused of practical inconsistency of commitments in advocating openness of discussions to public scrutiny. It is not hard to see how this argument is quite a powerful and effective use of dissociative reasoning.

Case 79

The *Spectator* in Hamilton, Ontario, has been running a steady stream of stories on the number of closed Board of Control and Board of Education meetings and the need to open them to the press and the public. The *Spectator* argues that those in public life hold a position of public trust and as such should have their debates and deliberations open to public scrutiny. But it does not believe those same rules should apply to itself.

The *Spectator* revealed in a front-page story last summer that Mayor Jack MacDonald had gone on a free fishing trip as a guest of Nordair, which operates flights out of Hamilton Civic Airport. The story continued to attract major attention in the news pages for a week (MacDonald eventually paid for the jaunt).

The irony of the situation, however, is this: for years, the *Spectator*'s travel writer roamed the world on free airline passes. The *Spectator* claimed it could not afford to pay his way. When a complaint was brought before the Ontario Press Council, the *Spectator* first succeeded in insisting the hearing be held behind closed doors. Then *Spectator* publisher John Muir successfully blocked the council from hearing testimony from Norman Isaacs, the continent's leading expert on press ethics. Mr. Isaacs is on the teaching staff of Columbia University's Graduate School of Journalism and is a distinguished American publisher and editor. He was at the time also adviser to the National News Council of the United States. Unfortunately, the stance and attitude taken by the *Spectator* is not peculiar to that paper.

The word 'irony' here relates to the incongruity between the editorial policy often expressed by the *Spectator* and the actions of its editor in blocking a debate to public scrutiny. This inconsistency of commitments raises serious questions about the credibility of the *Spectator*.

What is questionable in this case is whether news media organizations like the *Spectator* are "in public life" and "holding a position of public trust" in the same respect or way that elected officials in government organizations are. The media are in public life, but perhaps the nature of their being in a position of public trust is different, at least in some respects, from that of people in private sector organizations.

6. Rule-Based Reasoning

All institutions have rules that define expected ways of doing things and have been drawn up and tacitly or expressly agreed to by the participants. Argumentation from an established rule is employed where one participant tries to persuade another to act in a certain way, and the other participant is resisting this persuasion attempt. A typical type of case is the following.

Case 80

> **Student:** I can't get my assignment in by Friday. Can I have an extension?
> **Professor:** What is your reason?
> **Student:** I have too many other things to do.
> **Professor:** We all agreed at the beginning of the year that this assignment was to be in by Friday. That rule stands, for everyone, unless you have a medical excuse. It is up to you to organize your work load.

The professor could back up this argument further by saying that if one student is given more time, it would be an unfair advantage over the others. The universality of applicability of the rule is here appealed to.

The argumentation scheme for the argument from an established rule is the following.

> If carrying out types of actions including state of affairs A is the established rule for x, then (unless the case is an exception), x must carry out A.
> Carrying out types of actions including state of affairs A is the established rule for a.
> Therefore a must carry out A.

The following are the critical questions appropriate for this argumentation scheme.

1. Does the rule require carrying out types of actions that include A as an instance?
2. Are there other established rules that might conflict with, or override this one?
3. Is this case an exceptional one, that is, could there be extenuating circumstances or an excuse for noncompliance?

The third critical question can be extended into a refutational argumentation scheme opposed to the argument from an established rule. In the argument for an exceptional case, a pleader claims exemption using the following argumentation scheme.

> If the case of x is an exception, then the established rule can be waived in the case of x.
> The case of a is an exception.
> Therefore the established rule can be waived in the case of a.

This refutational argument throws a burden of proof back onto the side of the proponent of the argument from an established rule. For example, in relation to case 80, the student could present a note from a physician verifying illness as an excuse.

Argumentation from precedent can be a way of supporting the legitimacy of an excuse or even a way of arguing to refute or undermine an established rule.

> *Case 80a*
>
> **Student:** I heard you tell another student that he could have an extension of one week because his mother is ill. Well, my grandmother has not been feeling well. So I should be able to have an extension of one week too!

The argument from precedent is a species of case-based reasoning that uses argumentation from analogy between two cases. The principle, according to Golding (1984, 98) traces back to Aristotle's idea of justice as treating like cases alike.

The argumentation scheme for the argument from precedent is the following.

> Generally, according to the established rule, if x has property F, then x also has property G.
>
> In this legitimate case, a has F but does not have G.
>
> Therefore an exception to the rule must be recognized, and the rule appropriately modified or qualified.

The critical questions matching this argumentation scheme are the following.

1. Does the established rule really apply to this case?
2. Is the case cited legitimate, or can it be explained as only an apparent violation of the rule?
3. Can the case cited be dealt with under an already recognized category of exception that does not require a change in the rule?

The use of argumentation from precedent often poses a genuine puzzle or conflict that may need to be reasoned out by dialogue on both sides of the issue. It may lead to agreement on a new rule or to modification of the old rule.

Argumentation that appeals to personal sympathy or pity is often used in offering excuses or asking for leniency in enforcement of a rule. Such arguments do sometimes have a place. For example, they are recognized as having a place in judging sentencing for a crime or infraction of law. Very often, however, such arguments are rightly suspected as tactics to cover up a weak basis for an appeal or to try to distract from the real issue.

7. Position-to-Know Reasoning

Argumentation from position to know occurs in a dialogue where one party has access to knowledge the other party lacks, and the second party needs to act on a presumption or derive a provisional conclusion based on the first party's say-so. For example, a tourist in Winnipeg wants to go to the Convention Center and approaches a passerby on Portage Avenue.

Case 81

Tourist: Excuse me, could you tell me how to get to the Convention Center?

Passerby: Yes. Just go straight down Portage Avenue in this direction [indicates]. It is about eight blocks or so. Then turn to the right on Edmonton Street. It is only three blocks or so from Portage.

The tourist does not know how to get to the Convention Center, but she presumes that the passerby lives in Winnipeg, and would be likely to know. The positive response of the passerby confirms this assumption.

The argumentation scheme for the argument from position to know is the following structure.

a is in a position to know whether *A* is true or false.
a asserts that *A* is true (false).
Therefore *A* is true (false).

The critical questions appropriate for this argumentation scheme are the following.

1. Is *a* in a position to know whether *A* is true or false?
2. Is *a* an honest (trustworthy) source?
3. Did *a* really assert that *A* is true (false)?

Argumentation from testimony, for example, by an eyewitness in a court of law, is a species of argument from a position to know. The second critical question obviously relates to argumentum ad hominem and is generally permissible as a line of questioning in cross-examination of a witness in a court of law.

A special type of argumentation from position to know is argument from expert opinion, where one party draws a conclusion from advice or information given in dialogue by a second party who is an expert in a particular domain of knowledge. The argumentation scheme for the argument from expert opinion given in Walton (1989a, 193) is the following. *D* is a domain of knowledge or expert skill.

E is an expert in domain D.
E asserts that A is known to be true.
A is in D.
Therefore A may (plausibly) be taken as true.

The following five critical questions matching this argumentation scheme are given in Walton (1989, 194–97).

1. Is E a genuine expert in D?
2. Did E really assert A as true?
3. Is A relevant to domain D?
4. Is A consistent with what other experts in D say?
5. Is A consistent with known evidence in D?

As a basis for drawing a conclusion, appeals to expert opinion often tend to be questionable because the expert is not named, or is not really an expert at all, or was quoted wrongly, or was not even quoted at all. There are a lot of problems inherent in translating expert advice into layman's terms. Hence the argument from expert opinion is best treated as a presumptive kind of reasoning that shifts a burden of proof rather than deciding an issue conclusively.

Argumentation from ignorance is based on position-to-know reasoning. The argument from ignorance generally takes two forms: (1) A is not known to be true, therefore A is false, or (2) A is not known to be false, therefore A is true. A simple example is the following case.

Case 82
I do not know that there is a skunk in the cabin.
Therefore it is false that there is a skunk in the cabin.

The line of argumentation in this case is based on an implicit premise: if there were a skunk in the cabin, I would know it.

The argumentation scheme for the argument from ignorance is the following.

If A were true, then A would be known to be true.
It is not the case that A is known to be true.
Therefore A is not true.

This kind of argumentation can be stronger or weaker, depending on how conclusively verified the conditional in the first premise is. The basic principle behind the working of this premise is what de Cornulier (1988, 12) calls *epistemic closure*: "If it were true, I would know it." This type of counterfactual inference is a species of position-to-know reasoning.

A case from Walton (1992b) shows how epistemic closure can be stronger or weaker, depending on whether a knowledge base is complete, or epistemically closed, in a given case.

Case 83

The posted train schedule says that train 12 to Amsterdam stops at Haarlem and Amsterdam Central Station. It does not say that train 12 stops at Schipol.

Can we conclude then that train 12 does not stop at Schipol? It depends. If we know that railway policy is to always mark the name of every stop on the schedule, then we know that the knowledge base represented by the schedule is complete, or closed epistemically. We can infer that if there were additional stops, they would be posted on the schedule. Given the premise that it's not the case that a stop at Schipol is marked, we can conclude that it is not true that this train stops at Schipol.

Arguments from ignorance, often called lack-of-knowledge inferences in the social science literature, are often found in argumentation from expert opinion. For example, the following type of case is cited by Collins, Warnock, Aiello, and Miller (1975, 38).

Case 84

An expert is asked whether Guyana is a major producer of rubber or not. The expert knows that Peru and Columbia are major rubber producers and that if Guyana was too, she would be likely to be aware of it. She concludes: "I know enough that I am inclined to believe that Guyana is not a major producer of rubber."

The conclusion in this case is drawn on the basis of presumptive reasoning, because the expert does not know for sure that her knowledge base is closed (complete) with respect to major rubber producers in South America.

In the following case, Ted and Wilma are watching a televised movie on the Leona Helmsley story, the biography of a powerful woman in New York real estate who was convicted of income tax evasion.

Case 85

Wilma: Is Leona Helmsley still in jail?
Ted: Maybe she's still in there, because we'd probably hear about it if she got out.

In this case, the presumption is that Helmsley's release from imprisonment would be a newsworthy event, and therefore, if it were to happen, Ted and Wilma would likely hear about it. Since they have not heard anything about it, the conclusion can therefore be drawn by ad ignorantiam reasoning that she is probably still in jail. Or at least that is a reasonable presumption or guess on balance.

In this case neither Ted nor Wilma is an expert on the Helmsley case. Ted's conclusion is based, instead, on what could be called common knowledge, or things likely to be known to anyone who is reasonably well informed or who is in a position to likely be informed about such things.

The argument from ignorance is the fundamental, underlying basis of the concepts of presumption and burden of proof. Presumptive reasoning is only useful and appropriate where there are divided opinions on an issue, and the proposition on neither side is known (conclusively) to be true or false. Hence you could say that all presumptive argumentation is, in effect, argumentation from ignorance.

8. Source Indicators Reasoning

In this section, three common and important argumentation schemes are outlined. What they have in common is that they judge the plausibility of what was advocated by the source that advocated it. They share this characteristic with argument from position to know and its subspecies.

In ethotic argument (Brinton 1986, 248), the *ethos*, or character of the speaker, is used to transfer credibility (positively or negatively) to the proposition advocated by the speaker. This function derives from Aristotle's remarks in the *Rhetoric* and *Nicomachean Ethics*, to the effect that the good person's speech is more credible, especially where certainty is impossible and opinions are divided.

The argumentation scheme for ethotic argument is the following.

> If x is a person of good (bad) moral character, then what x says should be accepted as more plausible (rejected as less plausible).
> a is a person of good (bad) moral character.
> Therefore what x says should be accepted as more plausible (rejected as less plausible).

The critical questions for ethotic argument are the following.

1. Is a a person of good (bad) moral character?
2. Is character relevant in the dialogue?
3. Is the weight of presumption claimed strongly enough warranted by the evidence given?

Ethotic argument is often used to reduce or enhance the plausibility of a proposition for which other evidence has already been given.

Typically, ethotic argumentation is meant not to be conclusive by itself but to help tilt a burden of proof where other kinds of evidence are inconclusive or lacking.

Argument from bias is a refutational or dissociative type of argumentation used to attack the credibility of a source. Bias is hard to define, but a tentative definition has been advanced in Walton (1991b) that bias is a failure of critical doubt to function correctly in a dialogue that blocks openness to new or contrary evidence as the dialogue proceeds.

The argumentation scheme for the argument from bias is the following.

> If x is biased, then x is less likely to have taken the evidence on both sides into account in arriving at conclusion A.
> Arguer a is biased.
> Arguer a is less likely to have taken the evidence on both sides into account in arriving at conclusion A.

Whether bias is a bad or harmful thing very much depends on the type of dialogue the arguer is supposed to be engaged in. In a sales speech by a car salesman to sell his product to a potential buyer, a certain degree of bias is expected, without any deceptions being involved. We expect a different kind of argumentation, however, in an article in *Consumer Reports* where this type of car is evaluated. We also presume that *Consumer Reports* and its authors are not being paid by this particular car manufacturer.

The critical questions for the argument from bias are the following.

1. What type of dialogue are the speaker and hearer supposed to be engaged in?
2. What evidence has been given to prove that the speaker is biased?

Both ethotic argument and argument from bias are often based on innuendo, the creating of doubts or suspicions without sufficient evidence or when insufficient sources have been named.

In some cases, charges of bias can be clearly substantiated and justified.

Case 86

A speaker in a panel discussion on the issue of industrial pollution claimed to be a neutral scientist, but a critic showed that this scientist was on the board of directors of a large industry that had often been accused of pollution.

Here the concealment of the affiliation counts heavily against the claim of neutrality by the scientist.

In other cases it may be hard to prove that the alleged bias is a harmful bias. For example, in political debate, a certain degree of partisan advocacy of a particular point of view, for example, conservatism, is legitimate. If one participant in a debate accuses the other of a "right-wing bias," it may be hard to prove that this alleged bias should really detract from the credibility of this person's argumentation.

In the argument from popular opinion, the accepted practices or beliefs of a majority (often stated as "everyone," "nearly everyone," "all these people," etc.) are taken as a premise. The conclusion is that the hearer should also adopt the same policy or belief.

> *Case 87*
>
> Karen and Doug are tourists in a country that has lots of bicycle paths. Doug asks whether they should ride side by side or stay in single file. Karen replies: "Everyone else is riding side by side."

In this case, the argument seems reasonable. Karen is suggesting that riding side by side is an acceptable practice here, judging by what the others are doing. There is an element of "position to know" argumentation in this case as well, because Doug and Karen are tourists who are not familiar with the usual or accepted routines, whereas the other people, presumably, are in a position to know.

The argumentation scheme for the argument from popular opinion is the following.

> If a large majority of some reference group accept A as true, then there is a presumption in favor of A.
> A large majority of the reference group accept A as true.
> Therefore there is a presumption in favor of A.

This kind of argumentation is generally not highly reliable and is subject to qualifications and defeasibility, once firmer evidence comes in that may resolve the question more objectively. Argumentation from popular opinion is often a good tentative basis for prudent action, however, where an issue is open to divided opinions.

The critical questions for the argument from popular opinion are the following.

1. Does a large majority of the cited reference group accept A as true?
2. Is there other relevant evidence available that would support the assumption that A is not true?

3. What reason is there for thinking that the view of this large majority is likely to be right?

Critical question 3 tends to link the argument from popular opinion with other argumentation schemes like argument from position to know and ethotic argument.

Clearly argumentation from popular opinion is presumptive in nature. The argument form 'Everybody believes A, therefore A is true' is deductively invalid.

9. Practical Reasoning

Many argumentation schemes are species of practical reasoning and are used in connection with a particular aspect of practical reasoning. One of these schemes is that of argumentation from consequences, whereby a contemplated course of action is supported by citing its good consequences (positive form) or is rejected on the grounds that it will have bad consequences. Argumentation from consequences pertains to the side-effects premise in practical reasoning, whereby carrying out a goal is evaluated positively or negatively in relation to its cited good or bad consequences.

An example would be the following argumentation, used in a critical discussion on whether mandatory retirement is a good policy or not.

Case 88

One serious detriment of not having mandatory retirement is that millions of young people will be denied access to good jobs while senior citizens can go on and on performing inadequately into senility when they do not need the high income at that stage of their lives anyway.

This argument cites what are taken to be negative or bad consequences of adopting the policy of not having mandatory retirement. According to the argument, a weight of evidence is brought against this policy, thereby supporting the opposed thesis that mandatory retirement is a good policy.

The argumentation scheme for argument from consequences is the following.

If A is brought about, then good (bad) consequences will occur.
Therefore A should (not) be brought about.

Argumentation from consequences can occur in stronger or weaker forms, depending on the modal verb used in the conditional premise above. In the strongest form, the verb 'must' is used, while in the weakest form, the verb 'might' is used. In the middle version used above, the verb 'will' is used.

The critical questions for the argument from consequences are the following.

1. How strong is the likelihood that the cited consequences will (may, must) occur?

2. What evidence supports the claim that the cited consequences will (may, must) occur, and is it sufficient to support the strength of the claim adequately?

3. Are there other opposite consequences (bad as opposed to good, for example) that should be taken into account?

Typically, in policy discussions, positive argumentation from consequences is deployed against negative argumentation from consequences, as indicated by the third critical question.

Another argument used as part of practical reasoning is the argument from waste, or *argument du gaspillage*, described by Perelman and Olbrechts-Tyteca (1969, 279–81).

Case 89

The argument of waste consists in saying that, as one has already begun a task and made sacrifices which would be wasted if the enterprise were given up, one should continue in the same direction. This is the justification given by the banker who continues to lend to an insolvent debtor in the hope of getting him on his feet again in the long run. [279]

In the argument from waste, the proponent is trying to carry out a goal but encounters difficulties or costs, reasoning: "If I stop now, my previous efforts will have been wasted. Therefore I must continue." This is a species of negative argumentation from consequences, leading to a positive conclusion for action.

The argumentation scheme for the argument from waste is the following.

As a consequence of stopping trying to bring about *A*, the previous efforts will be a waste.

A waste is a bad thing.

Therefore continuing to try to bring about *A* is the indicated course of action.

This argumentation scheme clearly has a *modus tollens* structure as a species of practical reasoning.

The critical questions appropriate for the argument from waste are the following.

1. Is bringing about *A* possible?
2. Forgetting past losses that cannot be recouped, should a reassessment of the costs and benefits of trying to bring about *A* from this point in time be made?

A full analysis of this argument is not possible here. But suffice it to say that such an analysis involves the structure of cost/benefit decision making when calculating the likely costs or benefits of consequences.

A special subtype of negative argumentation from consequences occurs where the proponent indicates his willingness actually to bring about the bad consequences cited, and the respondent has reason to think that the proponent is capable of such action. The speech act, in such a dialogue, is a threat, and the argumentation involved could be called argument from a threat. Other kinds of argumentation as well use intimidation or appeals to fear (scare tactics) without a threat. Also, many threats are indirect speech acts, expressed overtly in the speech act of a warning (where no willingness to bring about the bad consequence is expressed overtly). We could call these covert or indirectly expressed threats.

The argumentation scheme for the argument from threat is the following, where 'I' designates the speaker (proponent) and 'you' designates the hearer (respondent).

> If you bring about *A*, some cited bad consequences, *B*, will follow.
> I am in a position to bring about *B*.
> I hereby assert that in fact I will see to it that *B* occurs if you bring about *A*.
> Therefore you had better not bring about *A*.

Argumentation from threat is incompatible with the goals of a critical discussion and is transparently inappropriate in a critical discussion. It is well recognized in the literature on negotiation, however, that threats, especially covert threats, can be allowed in some cases as legitimate bargaining moves. In this context, they are not prima facie fallacious as arguments.

Many other argumentation schemes are subspecies of practical reasoning or are closely related to practical reasoning. For example, many slippery slope arguments, as shown in the next section, are based largely, or even exclusively, on practical reasoning. Group policymaking discussions, for example, in political debates and deliberations, are typically sewn together as a fabric of practical reasoning.

Cost-benefit analysis is another type of argumentation of which we can make sense only as a species of practical reasoning. Legal reasoning, and inquiries into (or arguments based on) goals and intentions, are most often sequences of practical reasoning.

Thus practical reasoning is a general structure or type of reasoning that ties together many of the presumptive argumentation schemes and enables us to make sense of them as *linked* arguments, where both (or all) premises are needed to support the conclusion.

10. Gradualistic Reasoning

In gradualistic reasoning, called the *device of stages* by Perelman and Olbrechts-Tyteca (1969, 282), a sequential argument moves forward in a series of small steps rather than in one big leap. Instead of arguing directly from a premise P_0 to a conclusion C, the proponent may argue, for example, from P_0 to P_1 and then from P_2 to C and then, finally, from P_1 to P_2. The separation of the larger chain of argumentation into subarguments is a device to reduce strong opposition at any single point.

Slippery slope arguments are all based on gradualistic reasoning, but they can take four distinct forms, depending on how they use other argumentation schemes. The causal slippery slope argument is a species of argumentation from consequences of the negative type. A proponent of a causal slope argument warns a respondent: "Do not take this action you are contemplating, because it is the first step in a sequence of consequences that will ultimately lead to an outcome which is disastrous from your point of view!"

In the following example, an adult warns a young child to say 'no' to offers of drugs.

> *Case 90*
>
> Once you try even a supposedly harmless drug like marijuana, you might find that you are one of those people who becomes easily addicted and that you cannot stop taking it. From there, the progression to harder drugs like cocaine, and then even to heroin, is all too easy.

The argumentation scheme for the causal slippery slope argument (Walton 1992a, 93) is the following.

> A_0 is a contemplated course of action.
> Bringing about A_0 would likely have A_1 as a consequence, which would in turn lead to A_2, and so on, through a sequence to A_n.
> A_n is a horrible (disastrous) outcome.
> Therefore A_0 should not be brought about.

The critical questions matching this scheme are the following.

1. Is A_0 in fact the right description of the act being contemplated?
2. Do any of the causal links in the chain lack sufficient evidence to support the claim that they will (might, must) occur?
3. Is the outcome A_n as bad as suggested?

In many cases, the sequence of subarguments is only suggested by a kind of innuendo rather than filled in and proven. In such cases, a critic must try to get the proponent of the slope to be more explicit in filling in these steps. Otherwise, the weight of presumption in favor of the slope as an argument may be very small at best.

The precedent type of slippery slope argument combines argumentation from precedent with gradualistic reasoning in a slope format. In this type of slope, the proponent warns the respondent that a contemplated action will be the first step in a chain of precedents that will eventually lead to some horrible outcome.

> *Case 91*
>
> In an argument on prayers in the schools, a participant argued that if you accept one type of prayer as legitimate, that will set a precedent and, to allow for equal rights, you will have to accept other religious groups as having the right to introduce their prayers. Finally, this participant said, "It's a Pandora's box. You know that Satanism is a religion too!"

The argumentation for the precedent slippery slope argument is the following.

> Case C_0, once accepted, would set a precedent in place.
> The precedent of accepting C_0 would lead to accepting C_1 and so forth to C_n, where each step in the sequence would be bound to the prior one by case-to-case consistency.
> C_n would be horrible as an accepted practice.
> Therefore case C_0 should not be accepted in the first place.

The critical questions for the precedent slippery slope are the following.

1. Would C_0 set a precedent?
2. What is the evidence showing why each of the cited intervening sequence of precedents would occur?
3. Is C_n as intolerable as it is portrayed?

As with the causal slippery slope argument, the precedent slope is only as strong as the most weakly substantiated link in the sequence leading to the final outcome. A critic needs to search out the weakest links and challenge the argument there.

The verbal slippery slope argument combines argumentation from vagueness of a verbal criterion with gradualistic reasoning. It is a species of what was called above the argument of the beard and has often been identified with the "heap" or "bald man" argument of Eubulides. The following version is taken from Fogelin (1987, 72).

Case 92

If someone has one cent, he is not rich.
If someone is not rich, then giving him one cent will not make him rich.
Therefore no matter how many times you give a person a cent, he will not pass from being not rich to being rich.

This argument poses a paradox because both premises are true but the conclusion is false. Yet the argument seems to be valid, because the second premise can be applied over and over again to the first premise, in a sequence of modus ponens steps, in a way that seems to imply correctly that the conclusion is true. But this outcome is a contradiction. In a valid argument, it is impossible for the premises to be all true while the conclusion is false. Hence this type of argument is called the sorites paradox.

The argumentation scheme for the verbal slippery slope argument is the following.

Individual a_1 has property F.
If a_1 has F, then a_2 has F.
Property F is vague, and so generally, if a_i has F, then you can't deny that the next closely neighboring individual a_j in the series also has F.
Therefore a_n has F.
But quite clearly it is false that a_n has F.
Therefore you can't truly say that a_1 has F.

A fuller account of this argument, and the appropriate critical questions for it are given in Walton (1992a, chap. 2).

Finally, the all-in slippery slope argument combines features of all three previous types of slope arguments. It is typically used in arguments against euthanasia, for example, of the following kind.

Case 93

If we were to allow euthanasia in any form, even though it started out being "voluntary" for "terminally ill" old people, eventually it would become more and more widely acceptable as a practice. Soon it would be used to save expensive medical resources, and then gradually it would become a tool to eliminate retarded persons or those who are disabled. Eventually it would lead to a Nazi-like totalitarian tool, used to liquidate any perceived enemies of the state, like political dissidents.

This type of argumentation involves a causal sequence, precedents being set, the vagueness of terms like 'voluntary' and 'terminally ill,' and finally, a species of argument from popular opinion—once a practice becomes accepted, a social climate of opinion is put in place that makes it easier to move to the next stage.

A detailed analysis of the argumentation scheme and accompanying critical questions for the full slippery slope argument is given in Walton (1992a, chap. 5). One of the leading characteristics of the full slippery slope argument is that it incorporates a popular acceptance premise to the effect that once some new practice is started, it will gradually become more and more widely accepted. This factor often provides the moving force that makes the slope argument work. Thus the full slippery slope argument is related to the argument from popular opinion as well as to other subschemes. It is a complex type of argumentation.

6

Dialectical Relevance of Argumentation

The logic textbooks tend to appeal to the concept of relevance quite a bit in their classifications and evaluations of informal fallacies. But there appears to be very little, if any, consistency in how they do this. Nor is there any agreement among them on how relevance might be defined.

Copi (1982, 98) divides informal fallacies into two categories—fallacies of relevance and fallacies of ambiguity. Thirteen fallacies are classified as fallacies of relevance, including ad hominem, ad verecundiam, ad ignorantiam, ad misericordiam, and so forth. Copi does not try to define relevance, and this poses a theoretical problem for his treatment, because the thirteenth fallacy is *ignoratio elenchi* (irrelevant conclusion). This fallacy could generally be described as failure of relevance. According to Copi (110), "The fallacy of *ignoratio elenchi* is committed when an argument purporting to establish a particular conclusion is directed to proving a different conclusion." But if the other twelve fallacies are also failures of relevance, why have a thirteenth fallacy as a separate subcategory? This is a typical problem with the textbook treatments of relevance.

The plain fact is that relevance has never been defined in any way that could be useful in defining, analyzing, or classifying informal fallacies. Although formal relevance logics have been constructed, it has been shown in Walton (1982) that even the most potentially useful of these logics falls decisively short of capturing the concept of relevance appropriate for the study of fallacies.

What is needed is a pragmatic concept of relevance that does justice to the idea for any argument that there is a context of dialogue in which that argument was used as a contribution to the dialogue.[1] Thus the relevance of the argument, or any of its parts, is to be sought in understanding just how it or they contribute to the structure of the dialogue as a whole.

Now that we know some of the main types of dialogue, and some of the argumentation schemes used in them, we can usefully define relevance for the study of fallacies in two ways. First, argumentation schemes have appropriate critical questions attached—so a reply is relevant to an argument, at a local level, if it raises one of these appropriate critical questions. Second, arguments are judged in a global context of dialogue. Any argument, or other move in argumentation, is relevant to the extent that it fits into that type of dialogue as an appropriate move. Either of these types of relevance will be called *dialectical relevance.*

In this chapter, dialectical relevance is defined (briefly) as follows. A move in dialogue is dialectically relevant if it performs a legitimate function in the argumentation in some stage of a dialogue. Just about any argument or speech act in a dialogue could be "relevant" in some sense. But dialectical relevance, as defined here, is a normative concept that evaluates a move in argument in relation to the kind of argumentation that a participant in a dialogue is supposed to be engaged in, as appropriate for that type of dialogue. As shown in section 9 below, applying this definition requires an evaluation of six questions concerning what type of dialogue the participants are supposed to be engaged in, the stage of the dialogue we are in, the goal of this type of dialogue, the argumentation scheme for the type of argumentation involved, the prior sequence of argumentation, and the given institutional setting or social framework of the case.

Relevance is clear enough as a normative concept of reasoned dialogue, it will be argued in this chapter—a move is dialectically relevant if it is structurally coherent with the dialogue in which it occurs, meaning that it is a type of move that contributes in an appropriate way to the proper goal(s) of dialogue that the participants are supposed to be engaged in. But it will remain a substantive question to determine whether a particular move in an argument in a context of discourse in some specific instance of argumentation can be judged truly relevant in that case or not. The theoretical (structural) question can be clarified,[2] at least up to a point, but the practical question of judging relevance in a particular case is something else again.[3] There are many reasons for this gap.[4] The context of dialogue in a given case may be incomplete.[5] The participants may not have in fact clearly set out what each of their conclusions may rightly be taken to be. Or the argument may be in midstream (given the information available), and it may be hard to envisage what direction it appears to be taking.[6] For these reasons, it will become important to distinguish clearly between charges (criticisms) of irrelevance and fallacies of irrelevance.

In approaching the substantive practical question of how to determine relevance of argument moves in a particular case, therefore, in this chapter we will often take an oblique, negative route. Cases of criticisms of irrelevance, or at any rate, cases where there appears a strong presumption of irrelevance, and where that is a problem of dialogue, will be considered. By studying these cases, some appreciation can be gained of the kind and range of problematic argument junctures where irrelevance has become a cause for complaint or serious concern. Consideration of these practical problems of irrelevance in argumentation will go a long way toward helping us to understand how the concept of relevance in dialogue should function alongside the concept of fallacy. Ten case studies of criticisms of irrelevance will be evaluated.

1. A Classic Case Introduced

We begin with a classic case, that is, at the same time, one of the most difficult types of cases to analyze and to attempt to resolve in dialogue. Analysis of concepts and methods required to handle the problems posed by this type of case is taken up in section 7, after studying various simpler cases.

The case below comes from a political debate in the Canadian House of Commons on the Family Allowances Act (Bill C-70), pertaining to a motion to amend this act (Walton 1989b, 205–207). To understand the context of the case, the reader should know that the opposition had recently been attacking the government for allowing some cans of allegedly rotten tuna fish to be sold in supermarkets. The first speaker on the Family Allowances Act, Ms. Copps, related it to the "tuna fish scandal."

Case 94

Ms. Sheila Copps (Hamilton East): Mr. Speaker, I am not surprised that the Government has introduced the antifamily, antichild legislation which we see in the context of this particular proposed law, because we have seen quite clearly today that this same Government is prepared to play Russian roulette with the help of, potentially, a million Canadians. We heard the Minister of Fisheries and Oceans (Mr. Fraser) stand in his place in the House and say that the industry thought the standards were too high, so the problem was solved by lowering the standards. It is not a problem of aesthetics, Mr. Speaker, it is a problem of health. We are talking about a million tins of rotting tuna that the Government refuses to take off the shelves. I am not surprised, Mr. Speaker, that the Government would adopt a cavalier, devil-may-care attitude with regard to the issue of family allowances, child tax exemptions, and, indeed, the issue of missing children which I dealt with in my remarks yesterday. Quite clearly, a Prime Minister who can stand in his place, as he did today, and defend a Minister of Fisheries who has ignored—

The Acting Speaker (Mr. Paproski): Order, please. The Hon. Member knows that we are debating the amendment to the Bill on family allowances. I do not know why we are debating tuna fish. I hope the Hon. Member will get back on track.

The Acting Speaker's reply rightly questioned the relevance of the topic of tuna fish to the debate on the motion to amend the Family Allowances Act. The Acting Speaker has here advanced a criticism of irrelevance that seems to be a prima facie reasonable objection.

Ms. Copps responded to this objection in a very interesting reply, below, that did cite a connection of a sort between tuna fish and family allowances.

Case 94.1

Ms. Copps: With regard to the main question, which is family allowances, we are talking about the people who would be most affected by this cut-back. The Government says that $22 is not a lot for Canadian families. Well, families and single parents who are struggling to raise small children, often surviving on tuna, are being directly assaulted by the Government's anti-family budget measures. Look at the package. The Minister of National Health and Welfare (Mr. Epp) has said on numerous occasions that this particular measure will assist poor families.

According to Ms. Copps, families and single parents, "struggling to raise small children, often surviving on tuna," are being "assaulted" by the Family Allowances Act (called the "Government's anti-family budget measures" by Ms. Copps).

What has happened here? The Acting Speaker has challenged the relevance of the subject of tuna fish to the issue of the Family Allowances Act. In response, Ms. Copps has cited a connection, of a sort. But is it a serious enough connection of the right sort to rebut the criticism of irrelevance? Ms. Copps's argument can be outlined in the following schematic summary.

The government lowered standards for tuna fish, allowing millions of tons of rotten tuna to stay on the shelves.

This action showed a lack of concern for the people of Canada and demonstrated that the government was wrong.

Therefore the government lacks competence and concern for the people and is showing this same bad attitude once again with respect to the family allowance bill currently the subject of debate.

Therefore we can reject the government's argumentation for its motion on this bill.

We can see that there is a kind of relevance in this argumentation. Ms. Copps is bringing a kind of ad hominem attack to bear by linking what she takes to be the government's bad attitude or "lack of concern" on one issue with the same fault on another issue. But is this kind of relevance enough of a connection to sustain the claim that raising the tuna fish issue is dialectically relevant at this stage of the debate on Bill C-70 in the House of Commons?

It would seem not. The basic reason is that the particular issue of the tuna fish, even if it could be resolved, would seem to carry little or no weight in influencing anyone reasonably to vote for or against the Family Allowances Act. Questions of inspection or standards of quality of tuna fish might be issues of deep concern for the electorate and hotly contested topics of debate on certain occasions. But would a debate on a motion to amend the Family Allowances Act be such an occasion? Clearly not, unless some substantial connection could be established by someone who alleges relevance. The dialectical problem is that not every issue can usefully and reasonably be debated in a single session, where a motion must be decided.

What is really going on here is that the issue of tuna fish has been the subject of a recent scandal, and Ms. Copps finds it a convenient vehicle to continue an attack on government policies generally. She is not deterred by the Acting Speaker's request that she "get back on track" and insists on using the debate for her own purposes. The Acting Speaker's intervention failed to thwart her continued attack.

This case shows how difficult it can be for a moderator to enforce reasonable requirements of dialectical relevance in a debate on a particular issue. For an adventitious connection can always be found by a creative and determined debater, who may then claim relevance. This type of aggressive move may require both good judgment and strong action on the part of a moderator if she is to permit the issue to get a fair hearing from both sides of the argument. The problem is quite a serious practical difficulty.

Once an agenda and format of dialogue have been set and agreed to by the participants, at the opening and confrontation stages of a dialogue, it can be a serious problem if one participant decides unilaterally, and without the other's permission or assent, at the argumentation stage, suddenly to turn to discussing a different issue or even to begin engaging in a different type of dialogue. The problem, in such a case, is that such a shift will take time and attention away from the issue that is supposed to be resolved.

An agenda is not sacrosanct and can be changed, in some instances, once it has been set. But the right conditions are required to make such a change appropriate and legitimate. Especially once this side has agreed to a particular agenda and format, whether tacitly or explicitly, then objections of irrelevance should be judged as relative to such prior agreements.

This case suggests that dialectical relevance is a pragmatic matter. Two issues in a dialogue may be relevant but not dialectically relevant in a way that would justify a lengthy speech on one when the issue to be resolved on the agenda is the other. How to deal with what appears to be a tactic of digression that threatens to disrupt the original debate is the problem.

This problem is one of argumentation tactics, because each participant in a contentious argument will always try to bring the discussion around to focusing on the issues where his side is strongest, and the other side appears to be open to criticism, or on weak ground generally—see section 6.7 below on how such tactics work, in relation to relevance. Although judgments of the relevance of an argument can therefore be controversial, it does not follow that they cannot be adjudicated by appropriate and impartial critical standards.

Schopenhauer (1951, 29) clearly recognized this tactic and defined it in his list of dialectical stratagems. It is number 29 in Schopenhauer's list (quoted below).

> If you find that you are being worsted, you can make a *diversion*—that is, you can suddenly begin to talk of something else, as though it had a bearing on the matter in dispute, and afforded an argument against your opponent. This may be done without presumption if the diversion has, in fact, some general bearing on the matter; but it is a piece of impudence if it has nothing to do with the case and is only brought in by way of attacking your opponent.

Schopenhauer (29) offered the following example to show how this sophistical tactic works. Suppose two parties are having an argument, one maintaining that the government system in China is very good, because promotion is based on competitive examinations rather than on hereditary nobility. The other party tries to argue against this position by maintaining that education does not make a good criterion for judging people for suitability to hold a government office, but he appears to be losing the argument. What he could do, to try to turn the situation around, according to Schopenhauer, is to "make a diversion" by arguing, for example, that "in China all ranks are punished with the bastinado" (29). This emotional riposte makes the Chinese look somehow bad or guilty and therefore seems to count heavily against the contention that the Chinese system of government is good. Schopenhauer calls this the tactic of *diversion*, for the trick is to lead the opponent into an abandonment of the original point in dispute.

Dialectical relevance needs to be viewed, in a particular case, as defined relative to the goal an argument should have in its context of dialogue. Such goals should ideally be specified at the opening and confrontation stages of an argumentative discussion and then subsequently enforced, in order to support these goals. It is a question of what the original point of a dispute is supposed to be. Dialectical relevance is a normative concept, so defined, that sets a standard for argumentation that is reasonable in the practical sense of contributing to the goal of a type of dialogue the participants are supposed to be engaged in.

From a more practical point of view, however, strictness of standards of relevance needs to be interpreted in light of the institutional setting or other practical parameters of a discussion. In a business meeting, standards may be very strict, whereas in a philosophical discussion, standards of relevance could be much more loose and flexible. We must be aware of dialectical shifts as well. Strict adherence to the letter of commitments could be an appropriate standard for a legal dispute in court. But much more flexible and looser standards could be appropriate for a discussion that is supposed to be friendly and informal.

Thus theoretically, a subject can be relevant, but the practical implementation of guidelines in a particular discussion, in a given speech event, may require plenty of judgment and interpretation of the context. This much said, the general question can now be posed. Has Ms. Copps committed a fallacy of relevance? In line with the new pragmatic theory of fallacy, a case can be made out for saying that she has committed a fallacy of relevance, as follows.

Ms. Copps adopted the tactic of attacking the government on the tuna fish issue during the course of a debate on Bill C-70, a bill on a different issue. She persisted quite aggressively in this attack despite the procedural objections of the Acting Speaker. We really must look over the whole text of dialogue of the debate to reach a firm conclusion on how Ms. Copps's argument should be evaluated. But even from the parts quoted, several characteristics are clearly present that could raise the charge that she has committed the ignoratio elenchi fallacy. Though she may not have had deceitful or dishonorable intentions, Ms. Copps pressed forward so aggressively and persistently with tactics of attack on a fundamentally different topic of concern from the debate on the motion to amend Bill C-70 that it would be fair to judge that her technique of argumentation was used in such a way as to disrupt appropriate argumentation that would fulfill the goals of the dialogue by giving a fair discussion of the amendment motion on Bill C-70.

In the part of the debate quoted in case 94.1, Ms. Copps laid out the basis of her tactic by insisting that there is a legitimate connection between the tuna fish issue and Bill C-70. In following up this tactic, she then introduced a barrage of "facts" concerning Bill C-70 and then used an ad hominem argument to attack the government (Walton 1989b, 205–207). She accuses the government of supporting "friends in Bay Street" (the financial section), arguing that the government is against poor people and women, whom it oppresses. This is a shift to eristic dialogue, dividing or polarizing the dispute into two sides—the "good guys" (her party) and the "bad guys" (the government party).

Case 94.2

Sad to say, we do not see any leadership among Progressive Conservatives with respect to programs concerning women. Mr. Speaker, I must tell you that a great many women who are housebound are in no position to make personal representations to the Government to see renewed justice in terms of family allowances. So it is up to us in the Opposition parties to show the Government how strongly opposed the people are to the decision of the Minister of Finance (Mr. Wilson) and particularly the decision of the Prime Minister to slash $2,000 off the purchasing power of the poorest and most underprivileged families in Canada. I am talking about families of four persons who earn $15,000 on average. Those are not rich people. Bankers do not need the protection of the House of Commons any more than the big tuna industry, but it is quite obvious that the Minister of Fisheries and Oceans (Mr. Fraser) is prepared to speak up for companies and apparently could not care less about the health of thousands of Canadian men and women. The Government changed, quite clearly, in its approach concerning the promises made to the Canadian people.

This part of Ms. Copps's speech is relevant to the issue of Bill C-70, because it does relate to the alleged financial consequences of the bill. But clearly she has now turned the argument into an explicit ad hominem attack, alleging that the government party is part of an oppression of women and the poor, supported by the bankers and the "big tuna industry."

The general tactic used by Ms. Copps is to bulldoze right over the

Acting Speaker's ability to raise objections of irrelevance by mixing up the topics and bringing in "the big tuna industry." By this means, she can attack the government on the tuna issue with impunity, even launching into a major ad hominem attack on this basis, even though the aspects of the tuna fish issue that she brought up are not dialectically relevant to the present debate on the Bill C-70 amendment.

Debate on a bill in Parliament is limited to a fixed time, and therefore irrelevance should not be allowed by the Speaker if the irrelevant speech threatens to take up so much time that it would block the sequence of debate and prohibit other legitimate argumentation from being presented. If a speech seems to be heading into irrelevant matters, the Speaker should challenge the participant to "get back on track" or allow others to speak. But if the participant in question persists at length by discoursing on a topic of questionable relevance, beyond the bounds of propriety and good conduct appropriate for that stage of the debate, it may be helpful to bring forward the procedural objection that a fallacy of relevance has been committed. It is a question not of what the participant actually intended in so acting but of the tactics he or she used at that particular stage of a given dialogue. To the extent that the use of these techniques in this particular case can be said to be contrary to the goals of the dialogue, or not coherent with the proper sequence of the dialogue, we can say that a fallacy has been committed.

The other theoretically interesting aspect of this case is that it illustrates the problem of pinning down a charge of fallacy. The Acting Speaker tried to get Ms. Copps "back on track," but amazingly, she went so far as to retort that her attack on the tuna fish question was really relevant, because the "families and single parents who are struggling to raise small children" are "often surviving on tuna." This retort is patently ridiculous, even though it could quite well be true. But it puts the Acting Speaker in the position of having to try to show why it is ridiculous if he wants to back up his objection. This attempt to shift the burden of proof illicitly is clearly fallacious, because there is good textual evidence that Ms. Copps used a tactic of argumentation in such a way as to block and interfere with the constructive discussion of the bill supposed to be the subject of the debate and that she threatened to shift the debate to a quarrel.

2. Dialectical Relevance as a Pragmatic Concept

In a critical discussion, each participant has his or her proposition to be proved.[7] The pair of propositions to be proved on each side, when put together as a pair, define the *issue* to be resolved by the dialogue. The goal of each participant is to resolve the issue in his or her favor.[8] Any move of a type that fits into the proper normative structure used to facilitate this goal (relative to the rules of the dialogue) is relevant.

In a symmetrical type of persuasion dialogue called a *dispute,* the one proposition to be proved by the one side is the opposite (negation) of the proposition to be proved by the other side.[9] Some persuasion dialogues, however, are asymmetrical in the following way—the Proponent has to prove a particular proposition, whereas the Respondent need only show that such a proof is open to question in order to achieve his goal successfully in the dialogue.[10] This type of dialogue could be called a *weakly opposed dispute.*[11] Therefore, a relevant move on one side of an argument may not be relevant on the other side.

Burden of proof in persuasion dialogue is an allocation of the weight of evidence required for a participant successfully to persuade her opponent, that is, to prove the proposition she is set to prove as her goal in the dialogue.[12] A *presumption* is a speech act whereby a premise is granted or conceded to one's opposite number in a dialogue, even though this proposition has not been proved by the one who asks for it to be granted.[13] Presumption and burden of proof are devices that serve to shorten an argument and make its resolution practically possible even if the issue is one where lack of access to evidence makes a high standard of proof difficult or impossible.[14] Relevance is often judged in relation to presumptions.

The proposition to be proved by each side may be a set of propositions (a conjunctive proposition). If so, there will be a set of issues. Such a set is called an *agenda* of the dialogue. Sometimes the order of the propositions in the agenda does not matter, but in some cases it may be very important and even subject to dispute and negotiation prior to the opening of the dialogue. Judgments of relevance are always relative to prior specifications of propositions in an agenda.

Criticisms of irrelevance in dialogue may be brought forward by one participant if the other participant brings into play, at some move, a proposition not included in the agenda. As we will see, the situation is actually more complicated,[15] but basically it is the agenda of a dialogue that defines relevance internally, in a persuasion dialogue, and accounts for the relevance of argumentation in the dialogue at subsequent stages.

Evaluation of criticisms of irrelevance therefore presumes that an agenda exists and has been stated and agreed to by the participants in a dialogue. This assumption is not, however, always met in texts of argumentative discourse that have a dialogue format. In such cases, the criticism or question of irrelevance cannot be evaluated or resolved until the agenda is clarified or established. A problem here is that some arguments are not really about what they appear to be

Figure 2   Sequence of Dialogue

about. Hence, in some cases, criticisms of irrelevance cannot be decisively supported or refuted until further work is carried out. Several case studies of dialogues illustrating this kind of analysis are presented in Dascal (1977), Walton (1982), and Sanders (1987).

A dialogue is a sequence of individual pairs of moves (messages, speech acts) starting at a first move and directed toward a goal. In figure 2, M_i is an arbitrarily selected move, and M_o is the initial move. Because dialogue is sequential, two moves in a dialogue M_i and M_j may be said to be *indirectly relevant* to each other, where M_i is relevant to some intervening move, which is in turn relevant to M_j. Characteristically, when relevance is cited in an argument, indirect relevance is meant.

Two propositions (messages, speech acts) are *locally relevant* in a dialogue where they are related to each other at, or in the region of, some particular move. For example, a reply may be locally relevant to a question if the reply is related in some appropriate way to the question where the question and reply occurred at the same move. Or a question may be relevant to an immediately previous question, meaning that it was related somehow to the question of the prior move in the dialogue. Local relevance pertains to some particular move M_i in a sequence of dialogue.

By contrast, *global relevance* refers to the relationship between a speech act at some particular move and the goal of the dialogue. Two propositions are said to be globally relevant in a dialogue where one is a particular move M_i in a sequence of dialogue, and it is related to the proposition that describes or defines the outcome that is the goal of the dialogue. In a persuasion dialogue, global relevance pertains to the relationship between the proposition or speech act of a particular move and the proposition that is the agenda of the dialogue as a whole.

But what kind of relationship are we talking about? The relationship of *topical relevance* is defined globally by a set of topics, T_i, that define what the argument is about. At the local level, topical relevance is defined as subject-matter overlap. Two propositions are related in the sense of subject-matter overlap where some topics are shared by the common subject-matters of the two propositions.[16] A subject matter is a subset of the set of topics of the dialogue that is assigned to a particular proposition in the argument.[17]

The relationship of *probative relevance* holds when one proposition can be proved or disproved from another. For example, 'All Athenians are Greek and Socrates is Athenian' is probatively relevant to 'Socrates is Greek.' The proposition 'Socrates is Greek' is topically relevant but not probatively relevant to the proposition 'Plato is Greek.' There are many theoretical, disputed questions in logic about the relationship between probative relevance and topical relevance.[18]

In some contexts of dialogue, it is clear whether or not one proposition is relevant to another. But in some cases, determinations of relevance depend on judgment and sensitivity to the context of the dialogue.[19] For example, at some particular move M_i at the beginning of a dialogue, it may be difficult to judge whether a proposition advanced in a speech act in the dialogue may turn out to be relevant to the lines of argument that will be developed at later stages of the dialogue. In such cases, an arguer may ask it to be granted that he can show later how this particular point will turn out to be relevant to the issue. Judgment of such requests depends on goodwill or trust in an arguer's integrity or ability to stick to the point (Gricean maxims of communication). When questions or criticisms of irrelevance arise in dialogue, generally the burden of proof should be on the one who has initiated a line of argument to give reasons why it should be considered relevant if it is challenged.

Relevance can be defined semantically or pragmatically.[20] While semantic relevance is often important at the local level of dialogue, more often criticisms of irrelevance in argumentation are pragmatic in nature because they pertain to the global level of dialogue. At the local level especially, criticisms of irrelevance often have to do with the question-answer relationship.[21] If one participant in dialogue feels that his respondent has not answered the first participant's question and has been evasive, he may accuse the respondent's reply of being "irrelevant." In evaluating this type of case, care is needed in many instances.[22] If the question was itself open to criticism or was unreasonably aggressive, the respondent may have been fully justified in not answering it. In some cases, there may be an obligation on the part of a respondent to correct or question a question. In such cases, the questioner should not be allowed to badger the respondent by unfairly calling his corrective reply "evasive" or "irrelevant."

3. Contexts of Dialogue

In some cases, a global criticism of irrelevance stems from a context of dialogue set by firmly established rules of dialogue. In a criminal trial, the prosecuting attorney is set the burden of proving the guilt of the defendant of the alleged charge "beyond reasonable doubt."[23] Any line of argument not clearly germane to proving this conclusion can be questioned for relevance.

Case 95

The prosecuting attorney in a murder trial argues at length that murder is a horrible crime and exhibits the victim's bloody jacket, emphasizing many gruesome aspects of the victim's death.

One problem here is that while it may be true that murder is a horrible crime, and true that the victim's death was gruesome in this case, neither of these propositions proves the conclusion that the prosecuting attorney is supposed to establish, namely that the defendant in this case is guilty of the crime of murder.

The objection in this type of case may take a number of specific forms. It could be argued that the attorney is appealing to pity (ad misericordiam). It could be argued that the considerations adduced constitute a weak argument, lacking "probative relevance."[24] It could be argued that by concentrating on items like the bloody jacket and so forth, the attorney is omitting other premises that should be accorded more consideration. But all these objections relate to the underlying problem that the attorney's line of argument is of questionable global relevance in establishing the conclusion he is supposed to prove.

As will be shown subsequently, many of the major informal fallacies (and most notably, ad hominem, ad baculum, ad misericordiam, and ad populum) are fallacies because of irrelevance in many cases. But in other cases, they are fallacies even where the appeal is relevant. Thus it is somewhat misleading to classify them as fallacies of relevance. On the other hand, irrelevance is a large part of the problem and accounts for fallaciousness of these arguments quite commonly. For they are all powerful arguments and therefore tend to carry (undue) weight even when they are not relevant in a dialogue. Often too, this irrelevance is covered up by a shift in the context of dialogue. For an argument that is relevant in one type of dialogue may not be in another.

In other cases, however, like 94, the irrelevance is not due to a misuse of any particular argumentation scheme associated with a major fallacy. In this type of case, the fallacy is best classified as just being generally a failure of relevance.

One might note here also that the criticism of irrelevance tends to be of a defeasible nature. The prosecuting attorney could possibly defend his argument against the criticism in some cases. For example, suppose that the horror of the crime was connected to proving the guilt of the accused. Suppose, for example, that the attorney could show that the accused had suffered from horrible nightmares and personality disturbances after the time of the crime and that these disturbances could be shown to be consistent with the trauma caused by committing a horrible crime.

The possibility of showing such unanticipated lines of relevance in an argument means that judgment is needed on the part of a mediator, chairman, or judge in ruling on relevance. For example, a lawyer, in the face of an objection of irrelevance, may offer the judge a promissory note, claiming that if the court will wait a bit, he can show why his line of argument is relevant. At the closure of the trial, in hindsight, it will have been clear whether the line of argument did turn out to be relevant or not. But in medias res, it may be problematic to discern whether a line of argument should be judged relevant or irrelevant.

Because dialogue properly has a creative aspect, it may be a good policy not to restrict relevance too tightly in many cases. Generally, the burden should be to allow apparent irrelevance, only challenging at the point where there is evidence of conflict with the goal of dialogue. Each specific context of dialogue may be different, however. Hence judgment is often required to sort out objections of irrelevance.

In the parliamentary debate in case 94, the Speaker of the House has the job of judging whether arguments are relevant or not. In parliamentary debate, there is a Gricean presumption that the members' contributions are relevant. But if a participant's arguments seem to be getting "off the track" instead of contributing to the debate, the Speaker must try to ensure a useful debate by asking the member to show why her arguments are relevant to the issue or to the bill being debated. Rules of politeness apply. The Speaker will gently try to encourage the member to either show why her remarks are relevant or to desist from a line of argument that does not appear to be relevant. The Speaker must use skills of judgment in making calls of relevance. Irrelevance is a presumption, however, that can become stronger and stronger as an offender wanders away from the issue.

The expression "red herring" is often used to refer to an argument that is claimed not to be relevant. In some cases, however, the allegation made by this claim is not a decisive refutation of the argument so discounted. Rather it is a criticism that questions the relevance of a particular argument. The less than decisive nature of this type of criticism may be appropriate in many cases, because there may be genuine controversy concerning whether something is an issue or not in a particular case.

The following case is a quotation of part of a newsmagazine article on the jury acquittal of Bernhard Goetz in the shooting of four black youths on the New York subway in 1984 (Press, Johnson, and, Anello 1987, 20–21). The part quoted below (21) comes just after a paragraph that concludes that the jurors were predisposed toward Goetz during the trial.

Case 96

After the trial, the red herrings surfaced immediately. One ready bromide: had a black man shot four whites, he would have been convicted. Possibly, but there is contrary evidence. Last year a grand jury refused to indict a black man who had killed a "white punk" threatening subway passengers with his fists.

It might be recalled that in fact many of the public commentators on this case contended that race was a major issue. The *Newsweek* article, however, argued that the Goetz case "crossed class and color lines" because so many people in the United States have become crime victims, whether black or white, and so in this case, there was considerable controversy over the question of whether race was really an issue or not.

By calling the cited color reversal argument a red herring, the *Newsweek* article is questioning the relevance of this argument. Note that the rejection of the relevance of the argument is not total and absolute refutation, however. The article concedes that it could "possibly" be an argument with some weight. And indeed, it concedes this point emphatically by taking the trouble to follow up with a citation of some allegedly contrary evidence—a color-reversed case where the black defendant was acquitted.

Hence in this case, the use of the term "red herring" makes it clear that a disclaimer of relevance is being brought forward, but the disclaimer is not meant to be a decisive refutation of the relevance of the argument cited. Instead, it is a criticism that questions the relevance of the cited argument and thereby shifts the burden of showing relevance to the other side of the controversy.

A potentially problematic aspect of this particular case is the question of legal relevance *versus* relevance in the broader sense relating to the case as an issue of ethics or public opinion. Legal standards of relevance may be narrower than standards of relevance acceptable in extralegal contexts of dialogue. According to Ilbert (1960, 16), for reasons of policy and convenience, "the courts have excluded from consideration certain matters which have some bearing on the question to be decided." Such matters are judged legally irrelevant in court even if in a broader sense they are relevant. Even if the color reversal argument cited in case 96 above were judged legally irrelevant in the

Goetz trial, it is a separate question whether the argument may or may not be relevant in some extralegal context of dialogue. Shifts of context of dialogue, as noted above, can be critical to correctly evaluating the worth of a criticism of irrelevance.

The cases in this section bring out three lessons that will be put to good use in the next three sections. The first lesson is that much of the evidence required to evaluate a criticism of relevance may not be given explicitly in the text of discourse for a particular dialogue. Indeed, the dialogue itself may be incomplete at the time the criticism is advanced. For this reason, it is often said, with some justification, that relevance is "a matter of degree." More correctly, however, it should be said that criticisms of relevance are often based on presumptions and that evaluating them therefore involves judgment and implicature.

The second lesson is that because of the provisional and context-sensitive nature of evaluations of criticisms of relevance, we should distinguish between strong criticisms of irrelevance (refutations of arguments) and weak criticisms of argument that question relevance. The third lesson is that it may always be wise to be on the alert for unannounced dialectical shifts of the context of dialogue on a particular case.

4. Failure to Answer a Question

Attempts to construct relevance logics in the past have generally presumed that relevance is a relation of propositions in arguments (entities that are true or false). This essentially semantic point of view, however, is not adequate to studying fallacies of relevance and criticisms of irrelevance generally in argumentation. For relevance often has to do with speech acts that are important in argumentation but are not simply propositions or with other assertive kinds of moves. Questions are a case in point.

Replies to questions in interviews, political debates, and other contexts of conversational argumentation are frequently irrelevant. This irrelevance is often unobserved. But when it is noticed, it can frequently be a basis for criticisms of irrelevance. Not only can a reply to a question be judged irrelevant, but a question itself can sometimes be rightly judged irrelevant in a context of dialogue. What is at stake in such cases may not simply be global irrelevance. Often it is a localized matter of whether a particular reply is relevant in relation to a specific question that preceded it in the sequence of dialogue. On the other hand, fallacies of interrogation often involve the use of aggressive tactics of posing questions where a direct—or apparently most "relevant"—type of answer, would unfairly trap the respondent in a losing situation in the dialogue.

A frequent argument strategy in political debating is to ask a highly aggressive, loaded, and multiple question that the respondent cannot directly answer without incriminating or damaging his own position. The follow-up is to accuse the respondent of an evasive reply when he struggles to deal with the question by replying other than with a direct answer. Cases of the type below are common.

> *Case 97*
>
> **Questioner:** Can you assure the people of Canada that your party's short-sighted and disastrous economic policies will not continue to cause spiralling inflation, by taking immediate action to see to it that interest rates will go no higher tomorrow?
>
> **Respondent:** The rate of inflation has been no worse under our government than it was when your party was in power. In fact, it was higher when your party was in power.
>
> **Questioner:** That's irrelevant! You haven't answered my question.

This type of question-reply sequence is characteristic of a good many interchanges in the question period segment of the debates of the House of Commons of Canada, recorded in *Hansard*.[25] The purpose of the question period is to allow the opposition parties to request information or press for action from the governing party on the important questions of the day. Often, however, the questions are highly loaded, aggressive attacks on the governing party. Parliamentary rules in *Beauchesne*, the book of parliamentary rules of debate, forbid many kinds of questioning abuses. But because the rules are vague and subject to interpretation, the Speaker of the House often seems to be inactive or ineffective in controlling abuses.

In case 97 above, numerous aspects of the question are open to reasonable criticism. One problem is that in a relatively free economy in a democratic country, the governing party may not be in a position to see to it that interest rates are fixed in a specific day on a short-term basis. And the questioner, fully aware of this, may be less than sincere in making such a demand. The respondent may know this too, but since the debate is televised and recorded in written form in *Hansard*, both members of Parliament may be very conscious of how the interchange will appear to the public.

In this case, the respondent counters with an aggressive ad hominem attack. He alleges that inflation was no better when the questioner's party was in power and that it was even higher. The suggestion is that the questioner is inconsistent, and therefore perhaps even dishonest and insincere, in criticizing someone for causing escalating inflation when in fact his own party caused even worse inflation when it was in power. The respondent is replying, in effect, that his questioner had no right to ask this particular question, in view of his previous performance and position on the issue.

The respondent's reply in case 97 is a species of circumstantial ad hominem argumentation that was fully analyzed in chapter 5. In this case, the argument takes the following form.

> You argue that our party caused inflation by its "short-sighted and disastrous economic policies."
> But the rate of inflation was just as bad when your party was in power.
> This is hypocritical (inconsistent) because you decry in others something you practiced yourselves.
> Therefore your argument faulting our economic policies can be dismissed.

This argument is typical of the circumstantial type of ad hominem because it alleges a practical inconsistency of the "You do not practice what you preach" sort as a basis for refuting someone's argument.

This type of ad hominem attack is both common and extremely powerful as a type of argument in political debate.[26] Because it shifts the dialogue away from the issue itself, however, and onto the arguer's personal commitments, actions, or consistency of position, respondents to it will often rebut by claiming that the attack is irrelevant to the real issue. Such a rebuttal, as in the case above, often carries weight. For example, in case 97 above, the questioner's final rebuttal seems to score a good point. For it is quite true that the respondent did not answer the original question.

The real question to be posed in analyzing this case, however, is whether the respondent ought to have answered the question. The general principle of dialogue at stake in this type of case is the following rule: if the question is not reasonable, the respondent is not obliged to answer it. In fact this rule may be superseded by an even stronger rule: if the question is not reasonable, that is, is open to reasonable criticism as a fair question, then the respondent is obliged to criticize the question prior to, or instead of, answering it. The principle of dialogue enunciated here is that, in some cases, it may be reasonable, or even required, for a respondent who is asked a question to question the question. In these cases, a question may be an acceptable reply to a question.

On the other side, however, such a Socratic response may be attacked by the original questioner, who may accuse it of being "evasive," "irrelevant," and so forth. Therefore a problem is posed with respect to ruling when such nonanswering responses to questions are acceptable. The solution is to be sought in relation to understanding the reasonableness or acceptability of the original question. If the question can itself be criticized as unreasonable in line with the goals of a particular context of dialogue, then a case for not answering it can be made. The unreasonableness of the question is, of course, not the only basis for the acceptability of not answering a question. But it is one important type of basis in cases like the one above. This type of case is extremely common in political debate and no doubt in other contexts of argument as well.

Note that in this case, the reply given was topically relevant to the subject matter of the question. That was not the problem. The problem concerning the questioner is the failure of the reply to be an answer to the yes-no question seeking a commitment to action. The perceived failure is a local failure of matching between question and reply that is evidently a failure of neither topical relevance nor probative relevance.

It seems to me that the irrelevance cited should be regarded as global rather than local in nature, for two reasons. First, the text cited could be part of a larger discourse. But second, even if it is not part of an explicit text of surrounding dialogue, the given text needs to be expanded out into a fuller sequence once it is analyzed. For the question is a highly complex question with numerous presuppositions that can be adequately understood only as a dialogue sequence of questions and replies.

5. The Global Roots of Local Relevance

Manor (1982, 72) distinguishes between semantic and pragmatic relevance in studying the structure of question-reply dialogues. She characterizes semantic relevance as requiring that either the hearer should provide a direct answer to a question raised, or he should provide "an eliminative or corrective answer," one that is informative to the questioner. According to Manor, pragmatic relevance is a broader category that admits replies like the following sequence.

Case 98

Who ate the cake?
Go ask mommy.

Here the reply is broadly relevant in a pragmatic sense, because it directs the questioner to a source where she can presumably find the answer to her question.

It may initially seem that pragmatic relevance is always global, but this case shows that this is not so. In many instances, pragmatic relevance is evident at the local level, between a single question-reply pair of speech acts, but to be properly understood, the single pair needs to be expanded some ways into a wider context of the dialogue sequence. For example, the directive reply 'Go ask mommy' is relevant in this case if there is some reason for thinking that mommy might be a good source of reliable information on who ate the cake. In this case, then, probative relevance of a sort is involved, but it has to do with directing a respondent to an authoritative source of information. Exchanges like that in case 98 above involve directing a respondent to expert advice or to someone in a position to know something.

Many instances of question-reply dialogue have to do with goal-directed actions. The kinds of connections that determine relevance in these cases have to do with means-end connections between pairs of act descriptions distributed over a sequence of actions related to a goal. Such sequences in relation to ad hominem arguments have been studied in Walton (1985a), but they are important in many contexts of dialogue where all sorts of criticisms of irrelevance are advanced.

Very often background links in a sequence of implied actions need to be filled in, in order to understand relevance in these types of cases. An example adapted from Sanders (1987, 92) might be cited.

Case 99

   Q: How should I study for your midterm exam, Professor?
   A: Read the book.

In this discourse, there is no explicit subject-matter overlap between the question and the reply. Clearly, however, the reply is relevant. The connection implied is that there is a means-end connection between the action of reading the book and the goal of studying for the exam. Expanding the sequence of actions still further, the action of studying is a subgoal of a further goal of passing the exam. This sequence of goal-directed actions is understood by the questioner, the respondent, and those of us listening to the exchange. Hence the relevance of the reply to the question is evident to all of us. But it is not explicitly given as subject-matter overlap in the dialogue.

Criticisms of irrelevance that occur at a localized level of dialogue often take the form that a question posed has not been answered. The objection may be that while an answer has been given, it is not an answer to the specific question posed. Therefore, the contention is that the answer is not relevant.

> **Opposition Critic:** In view of recent cutbacks at the university, will the government give us assurance that it will protect women's positions from cuts?
>
> **Government Minister:** The government is concerned about any cutbacks that affect women's studies, and is discussing the matter with university representatives.
>
> **Opposition Critic:** My question wasn't about women's studies. It was about protection of women in all faculties at the university.

Here the criticism in the critic's second reply is one that pertains to the local level. It was the critic's specific question that the minister allegedly failed to answer. But even here, the global aspect is present in the background assumption that the original question was itself reasonable in the context of dialogue. But the presumption is that the question was reasonable and that the question itself was not in question. Hence we are right to focus on the criticism at the local level.

6. Hard Judgments of Global Relevance

In order for a meeting to pursue its goal in a practical manner, discussion of peripheral issues, even if they are legitimate and important issues in themselves, must be discouraged to some extent or ruled not relevant by the chairman. Judgment is often required in such rulings, however, for the rejected issue may be defended as "relevant."

Case 101

An emergency meeting of the Library Committee is called to discuss the proposal to close the university library on Sundays, for financial reasons and because there are few library users on Sundays. In the middle of the meeting, a student representative takes the floor and starts a vehement and lengthy argument for increase of student aid funding. The chairman of the meeting interjects, suggesting that this meeting is not the place for a general discussion of the question of student aid funding. The student objects, claiming that student funding is related to library closure because some students live in poor housing conditions due to lack of funds and cannot study at home. Therefore these students need to use the library to study even on Sunday.

This type of case is quite crucial for any analysis of the concept of relevance in argumentation to address. The problem is that the student could quite possibly make out a case that the question of student aid funding is connected, even if it is connected indirectly to the issue of library hours. Yet even so, the chairman of the meeting could also be justified in ruling that the connection is not substantial enough to spend much time in this particular meeting on the question of student aid funding.

Cases of this sort occur in political argumentation where a specific bill is being debated, and one participant in the debate launches into a general discussion of political problems in the country and attacks the opposition on a broad range of issues. When questioned whether his remarks are relevant to the specific bill under discussion, he may retort that they are "very relevant" and may be able to cite specific connections to support this contention. Even so, the Speaker (moderator) of the debate may be justified in urging him to confine his remarks to matters relevant to the bill at issue.

This type of case shows that it is not as easy or straightforward to enforce criticisms of irrelevance with clear justification where the participant accused of irrelevance is very determined to press hard against the moderator of the dialogue. In ruling on the reasonableness of these judgments, much depends on the purpose of the dialogue, on practical constraints on the length of the dialogue, and on the nature of the issue. In a philosophy seminar, for example, a wide-ranging discussion covering many explorations into distant topics may be tolerated or even encouraged. In case 101 above in the Library Committee meeting, however, action one way or the other may be required within a limited time, and therefore it may be necessary and beneficial for the chairman to be careful in confining discussion to the arguments that are strongly relevant to the outcome of the issue. It may be important, if the meeting is to fulfill its function, for the main arguments pro and con to get a good and fair hearing and discussion. And therefore the chairman's obligations include discouraging the wasting of too much meeting time on issues that he justifiably considers too peripheral. Therefore, while such judgments are important in facilitating good dialogue, they involve practical judgment and good skills of dialogue management on the part of the chairman.

In this case, the student has found a basis for establishing a genuine linkage between the issues of student aid funding and library hour extensions. But the question is whether this kind of relevance is significant enough to constitute dialectical relevance. Within the framework of the meeting, given its purpose and the accepted conventions necessary to contribute conditions for the realization of that purpose, is extensively discussing the problem of student aid funding dialectically relevant in the argumentation of this particular meeting? If not, then what are the reasons why not?

The basic reason can be sought in the goal of dialogue for the meeting, which is to discuss the issue of library hour extensions. This goal is a subgoal of the larger goal of reaching some agreement or unanimity as the outcome of the meeting, based on the reasoned arguments presented by various sides of the contentious issue. This subgoal is in turn a means of facilitating the ultimate goal of improving library services in relation to the other objectives and services of the university. The bottom line is that a decision, based on the reasons aired in the meeting, is required to take action to change the hours or preserve the status quo.

The issue of student aid funding is a highly contentious issue, and one, the chairman might reason, that could not possibly be resolved or usefully discussed in the framework of the present meeting on library hours. Practical constraints of intelligent planning of meetings within the decision-making structure of the university are behind this reasoning.

On the other hand, the problem of student aid funding is serious, free speech is an important principle, and it is important for university officials to take note of grievances or strong feelings expressed by students. This case is highly significant because it shows that although the concepts of relevance defined in section 2 above are involved, the operational use of these concepts in this case must be tempered with severe practical and social constraints. The good chairman must have not only the practical sense to see to it that the meeting is not wasted and that it does deal with the business at hand. He must also have the grace, skill, and good manners to keep under polite control distractions that could subvert the proper goals of the dialogue.

A good chairman could add this issue to the agenda, or even convene another meeting where the issue is the first to be discussed, if she felt that the issue was important for the committee to discuss. Thus the agenda, or goal of a dialogue, is not unalterable. Probably, in this case, the goal of the dialogue is seen quite differently from the student's point of view and from the committee chairman's.

But much depends, in a case like this, on the stage of the dialogue we are in and on the institutional framework within which goals and types of dialogue are formulated and on particular arguments that arise in this framework. This particular committee may exist for a stated purpose. This purpose itself is open to discussion, but not every particular time or place may be appropriate as a setting for such a discussion.

Here again, then, dialectical relevance is best judged as relative to a given stage of dialogue in which an agenda, framework, and type of dialogue may or may not already have been agreed to at some prior point in the dialogue. The problem here is for the chairman to determine when or whether it is constructive to change an agenda or pursue a line of argument, in light of the previous stages of argumentation.

This type of problem shows the difference of level between normative questions of the structure of reasoned dialogue in the abstract and practical questions of how to conduct a discussion to achieve a particular practical goal in decision making by dialogue. There are presumptions of relevance or irrelevance, as judged in a particular case, depending both on the global structure of a dialogue and on the local relationships that may be presumed to hold or not in the future sequence (conjectured) in a given context of dialogue. But translating these presumptions into a judgment about the strength of dialectical relevance required to sustain or rebut a criticism of irrelevance in a particular case requires a practical knowledge of the goals of a particular context of dialogue and acceptable means of facilitating these goals, as well as balancing them against other goals that may be operative in a particular case. This is a judgment of what is practically possible.

One possible resolution of this case is that the chairman could cite good reasons for ruling a protracted discussion of student funding concerns at this point as not dialectically relevant to the discussion of the meeting. These reasons could be associated with a judgment that, while there is relevance between the two issues, the dialectical relevance at this stage is not enough to sustain a protracted discussion of this particular issue, unless some more direct connection can be demonstrated (a defeasible presumption).

Another interesting sidelight on this problem is the observation that it could be a mistake for a chairman, speaker, or moderator of a debate to be overly analytical in giving reasons for his judgment to declare something irrelevant in a meeting. Overanalysis in itself could be a delay and hinder the progress of the meeting. While it is important, from a point of view of the metatheory of a dialogue, to understand reasons of this sort, it may be desirable for a discussion of them not to intrude too heavily into the sequence of dialogue itself.

7. Making a Big Issue of Something

Argumentation in a context of dialogue has a pragmatic character because it is generally a lengthy sequence of connected subarguments that moves toward a goal. This pragmatic character of argumentation means that considerations of relevance often have a tactical component. For there are often choices regarding what the proponent of an argument should emphasize as the most relevant parts of it, given that his argument itself may be a complex sequence of interlocking moves and subarguments. Relevance of a point or subissue may be contested, and while the proponent may choose to emphasize some parts of his argument as most relevant to deciding the issue, his opponent may see other parts of the argument as more relevant. Tactically speaking, each side will tend to try to emphasize as most relevant the parts of the argument where he thinks he is on the strongest ground and his opponent is on the weakest ground.

Allegations of fallacies of irrelevance do not always pertain to perceived attempts to change or deviate from the global issue of a dialogue. These criticisms can also relate to more localized levels of argumentation where an arguer adopts the tactic of switching the subject of the discussion to the subarguments where his case appears to be the strongest. This tactic can be used as a way of producing a deceptive appearance of winning the argument even where one's argument, from a more global perspective, is weak. This ploy could be called the tactic of making a big issue out of something in an argument. In fact, it is the sophistical tactic of making one issue appear big while ignoring other (perhaps more relevant) issues, hoping that your audience will overlook this shift of focus.

> *Case 102*
>
> A partisan leader wants to persuade his group to attack an airfield. Hoping to suppress discussion of whether the attack is likely to be successful, he tries to occupy the discussion with the question of who should lead the attack.

The tactic used by the leader in case 102 is to try to prevent anyone from bringing up the question of the feasibility of the attack, from asking the prior question of whether the attack itself is a good idea. Instead, he tries to focus argumentation on the issue of who should lead the attack, presuming (without defending) the thesis that the attack is a good idea.

In the legal context of cross-examination of an expert witness, Weber (1981, 308) has advocated the same type of tactic of turning an argument on a local issue where your side is strongest.

> *Case 103*
>
> By planning and preparation, you might try to push or persuade the opposing expert into narrowing the case to a single issue or to several determinative issues upon which you are strong and right. It is not always possible to turn a case on one controlling issue. But surprisingly often it is. And even more often it can be made to *appear* that the case turns on a single issue on which you are strong.[27]

In the context of advice-giving dialogue where opinions are solicited from an expert, this tactic of trying to push the expert into some areas and away from others would seem to be more like a blunder than a fallacy—a failure to conduct or utilize efficiently your appeal to expert opinion in argumentation. In this particular instance, however, it must be remembered that the legal system of cross-examination of witnesses (including expert witnesses) is based on the adversarial system. In a trial, each side has the obligation of trying to make the strongest possible case for his own client and, ipso facto, to try to defeat or weaken the case made by the opposing side. And it is true that persuasion dialogue, in general, has an adversarial element— each side has a burden of proof to support one's own argument and, if possible, to defeat or undermine the argument of the other side. To fulfill this burden unfairly could be a fallacy.

It may be legitimate, within reason, however, for lawyers in a trial to adopt tactics of narrowing a case to issues where they appear to be on the strongest grounds. This type of tactic of irrelevance is not always a fallacy in this context, wherever it is used. Each case must be evaluated on its merits, in its proper place in a context of dialogue.

Precisely this tactic was recognized and clearly described by Schopenhauer (1951, 25) in his list of dialectical stratagems for getting the best of it in a controversy. It is stratagem number 18 in Schopenhauer's list: "If you observe that your opponent has taken up a line of argument which will end in your defeat, you must not allow him to carry it to its conclusion, but interrupt the course of the dispute in time, or break it off altogether, or lead him away from the subject, and bring him to others" (1951, 25). Schopenhauer links this tactic with other tactics of relevance, by pointing out that it can be followed up by the tactic of *mutatio controversiae*, the trick of twisting the original dispute onto another issue on which you have the stronger case. Quite clearly, this tactic can be combined very nicely with the ad hominem attack and other emotional appeals designed to sidetrack the discussion of the real issue by making something else appear more urgent and overriding as a subject for debate.

Any judgment as to whether the use of this tactic is fallacious or not in a particular case, however, depends on the circumstances of the given case and also on the context of dialogue. For generally, and not unreasonably, participants in a dispute will try to concentrate on the subissue where they have the strongest case. Schopenhauer does not go into this question of evaluation per se, being content to describe the tactic and show how it works.

Of course, such tactics of shifting the issue are to be watched for, and can very usefully be identified and defended against, by countervailing tactics of argumentation. Generally, then, shifting to a local issue where your side is strongest—while it may be a sneaky tactic and open to criticism—is not necessarily fallacious but may often be. And certainly it is a kind of tactic that should be open to criticism once revealed.

Where such a tactic does become fallacious is in a kind of case where it is overdone to the point of radically shifting topics to arguments that are only tangentially relevant at best, even though they can be used as tactical clubs to try to hit the opposition decisively. Case 94 is, of course, the classic instance of this type of attack. The fallacy here arises from the tangential relevance of the tuna fish issue, escalated out of all proper proportion, in the discussion of a specific bill on family allowances. It is clearly being used as a tactic to beat the government into submission or at least into some sort of admission of guilt. This is the kind of aggressive intrusion into a supposedly serious parliamentary debate on a specific bill that should not be tolerated by the Speaker of the House. It is here used as a sophistical tactic of attack in a context of dialogue where it is inappropriate, and therefore the use of the term 'fallacy' is warranted.

The term 'fallacy' is indicated in case 94 because the tactic is to try to force closure of the discussion, and rejection of the bill, before even discussing the specific clauses in it or otherwise looking to the relevant considerations. Instead, Ms. Copps soars into a general condemnation of the government, twisting the debate around to an issue on which the government seems most vulnerable to attack.

8. The Classic Case Reconsidered

In case 94, the Acting Speaker expressed his "hope" that Ms. Copps would "get back on track." This interjection was not a refutation of Ms. Copps's argument on the topic of tuna fish as a fallacy of relevance. As a criticism, it should be construed in a milder way. It was more like a polite request to Ms. Copps to steer back toward the substantive issue of the debate, namely the Family Allowances Act. The interjection placed a burden of proof or explanation upon Ms. Copps either to "get back on track" or to justify her excursion into tuna fish if she could. In case 95, the evolving dynamics of a dialogue in midstream left possibilities of showing relevance open. So too, in case 94, the Acting Speaker's polite request leaves open latitudes but expresses a polite request for assurances.

In her response, Ms. Copps stuck to the tuna fish issue, taking the option of drawing it into the arena of debate, as she portrayed it. Following on from the section of debate quoted as case 94, Ms. Copps continued to speak against the government fiscal policies at considerable length, accusing members of the government party of supporting their wealthy friends at the expense of families at the low end of the income scale. So the Acting Speaker's interjection failed to stem Ms. Copps's verbal assault on government policies, including the subject of tuna fish and other matters, from continuing for a lengthy interval.

The failure of the Acting Speaker to enforce standards of relevance in this case raises some serious questions about the general purpose, function, and moderation of parliamentary debate as an institution. The purpose of this debate ostensibly was to discuss the Family Allowances Act. The goal of Ms. Copps, however, was evidently to attack the government on any terms and issues on which the government seemed vulnerable. She was clearly more concerned with the attack itself than with an attack on a particular issue and covers health and finance generally, as well as a range of family-oriented issues, not to mention tuna fish. Ms. Copps's conclusion would generally seem to be that the government is not competent or trustworthy to serve the people of Canada. This is an instance of an attack on the integrity of the government party of a type associated with the argumentum ad hominem. What is indicated is a kind of dialectical shift from the persuasion dialogue to the level of the eristic dialogue or quarrel where direct assault on the morality and honesty of the opposing side of the debate becomes the uppermost objective of the attacking party. A similar type of case in this respect, called the Sportsman's Rejoinder, was studied as case 2. Here too a genuine connection between two issues was cited by an attacker, but it is not substantial enough to justify the force of the ad hominem attack. And indeed, the ad hominem attack often turns out to be a poor argument because it is a species of failure of relevance.

Among the many possible good arguments for or against the Family Allowances Act, the tuna fish scandal, aside from its value as an ad hominem attack, would seem to be a minor argument at best. Government fiscal policies, in the subsequent attack launched by Ms. Copps on a broad front, have also not been shown by her to be more than minor considerations in any serious decision to vote for or against the Family Allowances Act. Therefore, any sober assessment of the probative relevance of these other issues to the act would have to rate them as tangential at best. It seems, then, that the real raison d'être for introducing these issues is the attack itself. The genuine but minimal connection between the act and the subject of tuna fish is therefore really a rationalization for the attack.

If such interruptions are as freely allowed in parliamentary debate as this case suggests, the question is raised whether such debate as it exists is a rational way of reaching conclusions on government policies for action. Is parliamentary debate a real example of reasonable persuasion dialogue on the issues it purports to address? Or is it really a form of interest-based bargaining, a cut-and-thrust process of attack and defense to extract votes from those among the electorate who see it or read about it in the media reports? Is it a kind of spectacle of diversion where irrelevance is not only tolerated on a broad scale but is successful and rewarded as an approved mode of rhetoric? The worry is that if political decisions are not being made on the basis of open parliamentary debate that considers the reasons on both sides of the issue, then they are being reached by some other process. This other process is likely to be some form of interest-based bargaining among the dominant interest groups, conducted in private negotiations.

How harmful irrelevance in fact is, in a particular dialogue, depends on the purpose and setting of the dialogue and on practical constraints. If each side has an allotted time to present its side of an issue, then a side that wastes its time on irrelevant arguments is simply weakening its own arguments. In this context, irrelevance seems less like a harmful fallacy than an instance of poor strategy in arguing, more harmful to yourself than to anyone else in the argument.

But debates on a proposed piece of legislation in the House of Commons are more than purely private arguments between two parties. The outcome decides an important matter of public policy. It is an ideal, or at least a hope, of advocates of parliamentary democracy that the best, strongest, and most relevant arguments on both sides of the issue will be advanced and tested against each other in the arena of debate, so that the important considerations on both sides will be aired. As noted at the end of chapter 4, political debate is supposed to have at least some elements of a critical discussion. The use of blocking tactics to evade the issue can prevent such a discussion from happening. Therefore irrelevance, in some contexts of debate, can be fallacious. Of course, irrelevance can be open to criticisms or objections in other ways as well. But its obstructive consequences pose a special danger in political debates, where in especially severe cases, it deserves to be baptized as a fallacy in its own right under the heading of dialectical irrelevance.

9. The Wastebasket Category

In many of the cases studied in this chapter, the tactic of argument judged irrelevant also made use of appeal to pity or other emotions, contained an ad hominem attack, or involved questionable tactics used in posing or replying to questions. Does it still make sense, then, to have a general fallacy of ignoratio elenchi apart from these other fallacies? And does it make sense to follow the tradition of classifying these other fallacies under the general heading of fallacies of relevance? Surely we can't have it both ways. Either ignoratio elenchi is a more general term that is used to classify a whole category of fallacies, or it is a fallacy in its own right, on a level with these other fallacies.

It is a common theme of the textbooks that many fallacies like the ad hominem, ad verecundiam, ad baculum, and ad misericordiam come under the general category of fallacies of relevance. But are these three species of failures of relevance fallacies in their own right or merely particular instances of the broader fallacy of irrelevance (ignoratio elenchi) in argument?

Castell (1935, 23) took the point of view that a group of these traditional fallacies are special instances of a larger fallacy called "Irrelevant Evidence," where facts presented as grounds for a claim are irrelevant to that claim. Castell included four of the traditional fallacies under this heading. Curiously, he made no mention whatever of the argumentum ad baculum.

> The general term Irrelevant Evidence covers a multitude of logical sins: evidence may be irrelevant in various ways and for various reasons. Some special types of this general fallacy have been singled out and given names. These include the Argumentum ad Populum, the Argumentum ad Verecundiam, the Argumentum ad Misericordiam, and the Argumentum ad Hominem. Each of these famous old terms, although names merely of special instances of Irrelevant Evidence, will repay separate consideration. [Castell 1935, 23]

Is Castell claiming that the "famous old terms" he lists above are fallacious in their own right or that they are fallacious only insofar as they are failures of relevance (instances of the so-called fallacy of Irrelevant Evidence)? Although he seems, in the main, to be opting for the latter point of view, nevertheless it is possible to perceive the elements of a compromise in his approach. He does think that even though these "famous old terms" are names "merely of special instances of Irrelevant Evidence," they are individually important enough to merit separate treatment in a logic text. And in his own text, Castell does treat each of them individually.

This sort of compromise will turn out to be supported by the new pragmatic theory, because relevance has to be judged at both a global and a local level. The argumentum ad hominem, argumentum ad verecundiam, and so forth have been shown to have special argumentation schemes of which they are instances. Charges of a fallacy of relevance have to be judged both globally, in relation to the dialogue as a whole, and locally, in relation to the use of a particular argumentation scheme.

Relevance in argumentation can be determined only by appealing to a normative model of dialogue appropriate as the context for the given speech event, the actual segment of discourse presented in a case, which yields the evidence that argumentation of some particular type is taking place. For example, given a case of an argument, it may be established that the argument is, let's say, an instance of a persuasion dialogue as opposed to a negotiation. The normative model determines the goal of the dialogue, and this, in turn, determines the relevance of any particular speech act in the discussion.

Relevance depends not only on the normative model and global goal of a dialogue, however. It also depends, at the local level, on the nature of the individual speech act. If the speech act is a question, for example, then certain responses will, or will not, be relevant. For example, an answer might be relevant. Or a speech act other than an answer, like a reply that criticizes the question's presuppositions, could be relevant. If the speech act is an argument rather than a question, relevance depends on the nature of the argumentation scheme that is appropriate for this particular kind of argument.

Or to cite another example, suppose the argument is an appeal to expert opinion in order to settle a disputed question. Then a response questioning the qualifications of the expert (who was earlier cited) could be a relevant response for that argument. But apart from being paired with this particular type of argument, such a response could be irrelevant in a discussion.

Thus relevance can depend on the nature of an argumentation scheme at the local level of dialogue. Yet at the same time, the relevance of a whole sequence of moves in a discussion can be judged globally only by testing it in relation to the larger context of dialogue and, in particular, to the goal of the dialogue. If the major problem is one more of the use of any argumentation technique that goes contrary to the goals of dialogue at the global level, the label *ignoratio elenchi* is appropriate. If the major problem is the abuse of a specific argumentation scheme (e.g., ad hominem), then a specific subcategory of fallacy (like ad hominem) is appropriate, even if irrelevance may be involved, to some extent.[28]

In making a critical judgment whether an argument or other speech act (like a question or reply to a question) is dialectically relevant in the normative sense advanced in this chapter, a critic should look at the evidence given in a particular case. In particular, six kinds of factors need to be taken into account.

1. *Type of Dialogue* What is the type of dialogue the participants are supposed to be engaged in? If it is a critical discussion, then the argument or move in question should be judged as relevant or not, in relation to that type of dialogue. An argument that is relevant in a critical discussion might not be relevant, for example, if the dialogue is supposed to be an inquiry.

2. *Stage of Dialogue* A speech act that was relevant at the confrontation stage of a dialogue, for example, may be irrelevant at the argumentation stage.

3. *Goal of Dialogue* Relevance is always determined in relation to the goal of a dialogue. If the given dialogue is supposed to be a critical discussion to resolve a conflict of opinions between two opposed points of view, P_1 and P_2, then a subargument will be relevant insofar as it bears upon, or is related to, the resolution of the question of which is the stronger presumption, P_1 or P_2.

4. *Argumentation Scheme* But how is the subargument related to some issue of a dialogue, like the opposition between two propositions P_1 and P_2? It depends on the type of argumentation scheme for that subargument. For example, if the subargument is an appeal to expert opinion, then whether that subargument is relevant depends on its argumentation scheme. And if a reply to it is to be judged relevant or irrelevant, the judgment depends on the types of critical questions that are appropriate for that argumentation scheme. For example, the reply, 'Is the authority you cited really an expert?' would be relevant.

5. *Prior Sequence of Argumentation* Whether a subargument is dialectically relevant in an ongoing dialogue may depend very much on what sequences of argumentation have gone before in the dialogue. Any textual evidence of the prior sequence of argumentation in a dialogue, in a given case, is an important source of evidence in judging relevance of a new line of argumentation.

6. *Speech Event* The given institutional setting or particular speech event may impose constraints and special rules that help to define relevance in a given case. For example, if the argument is taking place in a legal trial, specific legal rules will help to define kinds of moves that are judged to be relevant or irrelevant for that type of speech event. Or to take another example, if the speech event is a committee meeting of some particular corporation or group, the rules and practical requirements of the group may impose all kinds of constraints on what kinds of speech acts are to be judged relevant

in a meeting. The chairperson will have to interpret these rules and practical constraints. Thus what is considered relevant in this setting might be quite different from the setting of an everyday conversation or discussion outside of such a setting.

Semantic relevance of propositions has been studied in Walton (1982). But the kind of relevance that is most generally useful in evaluating allegations of irrelevance, in relation to sophistical tactics fallacies like the ad hominem or ad verecundiam, is pragmatic relevance. Pragmatic relevance refers to relevance of a speech act in a larger context of dialogue. Pragmatic relevance is, therefore, equivalent to dialectical relevance. To use a big word, it could be called pragma-dialectical relevance. But judgments of pragmatic relevance also have critical import for evaluating cases of ad hominem, ad verecundiam, and other techniques of argumentation as reasonable or fallacious. Therefore, any judgment of pragmatic relevance involves both a postulation of an appropriate normative model of reasonable dialogue and an identification of the appropriate argumentation schemes that are supposed to be applicable to the case cited.

Generally, argumentation is a pragmatically organized sequence of connected speech acts that proceeds from an opening phase to a closing phase. The goal of the argument defines successful culmination, or closure. Therefore, whether a move in argument is relevant or not also depends on the stage of argument we are in. Pragmatically speaking, an argument is a line or sequence that begins at a starting point and progresses toward a goal or ideal point of closure. Any line of argumentation that deviates from this ideal line of goal-seeking, connected speech acts is, by definition, irrelevant (or better, it is dialectically irrelevant in that context of dialogue). An irrelevant line of argument is open to critical challenge, or questioning if it shows evidence of deviating from the line of argument.

Not all failures of relevance can properly be classified as fallacies of irrelevance, however. If a speech act or line argument appears to be wandering away from the main line of argumentation in a dialogue, it should be challenged, and the question should be raised as to whether it is relevant. If the move in question is so aggressive or obstructive that it blocks off the line of argumentation altogether, however—as in a filibuster, which prevents all further relevant dialogue on an issue—then it may be judged as an instance of a fallacy of irrelevance.

Still, a clear distinction between criticisms of irrelevance and fallacies of irrelevance requires a deeper analysis of the concept of a fallacy than has previously been available. This in-depth analysis of the concept of fallacy has been provided by the new pragmatic theory.

Once particular fallacies can be identified and understood as characteristic types of argument tactics that involve the abuse of certain argumentation schemes, it will become clear how tactics of irrelevance can be used as sophistical tools. And it will also become clear how they can be defended against.

Currently Frans van Eemeren, Rob Grootendorst, and I are conducting a collaborative research project, 'Dialectical Relevance in Argumentative Discourse.' It is the aim of this project to present a theoretical analysis of the concept of dialectical relevance in argumentative discourse that will be useful for those working in the field of argumentation. Clearly, more theoretical work needs to be done in order to construct a precise theory of relevance that would be useful to aid our understanding of the fallacies and to support criticisms of irrelevance decisively. On the other hand, studies of specific fallacies like the ad hominem, ad verecundiam, and ad ignorantiam will also contribute greatly to our knowledge of how relevance works as a category of argument evaluation and criticism, when specific tactics that have been employed are the main focus in evaluating a charge of fallacy.

10. The Importance of Relevance for Fallacy Theory

Relevance is often dismissed as being hopelessly vague and so forth and therefore of no use in the study of argumentation. We have seen that there are systematic reasons why this type of remark is based on something true. Five different kinds of evidence are required to substantiate a judgment of relevance in a particular case. Yet in the short examples of fallacies so often used by the textbooks, not enough context of dialogue may be given so that a critic can justifiably say whether the argument is relevant or not. Yet because judgment and conditional presumptions are often involved in criticisms of relevance, it does not follow that the concept of relevance is not useful. Such judgments are often conditional on what is known.

The basic idea behind the critical discussion as a normative model of reasoned argumentation is that an argument should be generally restricted, once the argumentation stage has been reached, to a particular issue, or set of topics to be discussed. And in fact we could never correctly evaluate criticisms of irrelevance in texts of argumentative discourse as justified or unjustified without appeal to this basic assumption. It is true that defining the agenda can be one of the most powerful tools for controlling a dialogue like a debate, critical discussion, or negotiation. And it is true that, in some cases, an agenda can and should be changed. But the basic idea defining dialectical irrelevance is that not every time, place, or stage of a dialogue is the appropriate point to introduce shifts. Without a systematic understanding of this idea within the theory of dialogue, we could never hope to understand criticisms like the ad hominem and ad misericordiam, and so forth, as fallacies that are an important part of the curriculum in logic and argumentation studies. Hence relevance is an important subject eminently worth further study in developing a critical concept of fallacy as a part of the field of informal logic.

Why is probative relevance important in a critical discussion? For it may seem harmless enough for an arguer to spend some time considering weak arguments. After all, can't he go on to take up the stronger arguments for his thesis, having finished with the weaker arguments? But the problem is that, in a persuasion dialogue, each side takes turns, and the number of moves is finite or limited. The strategy of each participant is not only to bring the strongest arguments to bear on your own thesis but to prevent your opponent from doing so or to refute or weaken his arguments. Hence part of this strategy is to fill the discussion with arguments where your side is strongest and your opponent's side is weakest. By dominating the discussion with these arguments, your aim should be to minimize the possibility that your opponent might introduce or concentrate on arguments where his side is stronger. Therefore, probative relevance is strategically important in persuasion dialogues.

In a good persuasion dialogue, the strongest arguments for both sides should emerge. If one side manages to confine the discussion largely to arguments where his own side is strongest, while excluding the arguments where the other side is stronger, then that strategy may be effective in winning the contest. But it may also be a disservice to facilitating a good discussion that presents both sides of the issue. Hence a failure of probative relevance may not only be a weakness of strategy. It may be a weakness that undermines the goal of persuasive dialogue. For this reason, such a failure should be open to criticism.

Failures of global irrelevance represent a discontinuity of a dialogue in the argumentation stage as it relates back to the confrontation stage. Local irrelevance is important mainly because of its global consequences in allowing a dialogue to get "off track."

Many specific fallacies of argumentation turn out to be due, at least to some extent, to failures of harmful irrelevance. Personal attack (ad hominem) is the most notorious of these fallacies and the most common diversionary tactic in political debate. Personal attack is the subject of chapter 7, section 5. It will turn out that relevance is an important aspect of ad hominem as a fallacy and indeed that we could not give an analysis of ad hominem as a fallacy without some prior understanding of the concept of relevance in argumentation.

But failure of relevance is not the whole story of the ad hominem. And much the same thing, it will turn out, is true of the ad verecundiam and the ad ignorantiam as fallacies. These three fallacies will turn out to be "fallacies of relevance" only in a partial way—as fallacies they are often failures of relevance but turn out to be not purely and exclusively failures of relevance.

What has been shown by chapter 6 is that there definitely is a need for a separate fallacy of ignoratio elenchi for fallacious use of irrelevant moves in argumentation. Where the main problem is irrelevance rather than some more specific fault coming under the head of another fallacy like ad hominem, it is useful to have some name for the fault.

Is fallaciousness just the opposite of relevance? No—because of the distinction between a criticism and a fallacy. A move in dialogue is irrelevant (not relevant) if it doesn't fit into the proper sequence of dialogue. If a move fails to be coherent, or fails to fit into a dialogue, a criticism of irrelevance is appropriate. But such a failure should not, in general, be equated with the fallacy of ignoratio elenchi. To be fallacious, the use of a technique of argumentation must be a strong kind of failure of relevance that shows the use of a tactic to subvert the goals of dialogue, or even block off the dialogue, or shift illicitly to another type of dialogue altogether. In case 94.1, for example, it was judged that a fallacy of relevance was committed because the perpetrator persistently and aggressively used tactics to divert and block off the proper subject of discussion from continuing.

Broadly speaking, there are two kinds of dialectical irrelevance, and neither of them is inherently or necessarily fallacious. One kind of irrelevance is *internal irrelevance,* where there has been an internal shift—that is, a shift within a dialogue—of either the topical or the probative sort. The other kind of irrelevance is *external irrelevance,* where there has been a shift from one type of dialogue to another type of dialogue. For example, if an argument started within a critical discussion but then shifted to a quarrel, the argument within the quarrel context could be quite irrelevant to the previous line of argumentation in the context of the critical discussion. This is an external shift, and therefore it is a case of external irrelevance.

But not all dialectical shifts are illicit shifts, and not all irrelevant arguments are fallacious. Irrelevance is a deflection or deviation away from an original issue and context of dialogue. Such a deviation becomes identical only with a fallacious use of argumentation when it is so serious that it blocks, or is incoherent with, the goals of the original type of dialogue the participants were supposed to be engaged in.

In many cases, the participants have simply not decided what type of dialogue they are supposed to be engaged in or what the agenda is supposed to be. In such cases, whether a speech act is dialectically relevant or not cannot be determined from the given information. In such cases, indeed, the participants may begin to argue about these procedural matters.

But in cases where there is evidence to show that the participants are supposed to be engaged in a particular type of dialogue and have agreed on the agenda, it is possible to reach a reasoned judgment of the dialectical relevance of an argument on the basis of the given text of discourse.

7

A New Approach to Fallacies

Now that we have the argumentation schemes, the typography of dialogue, and the concept of relevance as tools of analysis, we can go
ahead and use these tools to analyze the fallacies presented in chapters 2 and 3. In fact, it is already highly evident to the reader that
there is a very close matching or resemblance between the various
argumentation schemes of chapter 5 and many of the various informal fallacies presented in chapter 2. Since the argumentation
schemes represent correct (reasonable) forms of presumptive reasoning, and the fallacies represent fallacies or incorrect forms of the
same kinds of reasoning, all we need to do is to compare one with the
other. Then we can see how each fallacy is a type of argumentation
that was used wrongly and identify the failure of reasoning.

In fact, this will be a part of the research program for the analysis
of fallacies we will advocate and implement. But before we can even
get started, there are three general problems that need to be discussed. The first problem is that, while some of the fallacies correspond to particular argumentation schemes, others do not correspond
to any single argumentation scheme at all. These latter fallacies, at
least many of them, will turn out to be counterexamples to the hypothesis of equating the concept of fallacy to a single argumentation
scheme used wrongly. The second problem concerns the names of the
fallacies. Some of them imply that the type of argumentation described is always fallacious, or incorrect, while others appear to describe types of argumentation that could be correct, at least in some

instances. This is quite a general problem that seems to preclude any simple or uniform methodological approach to evaluating arguments said to be fallacious. And it leads to a third problem of how to identify the various fallacies. The textbook accounts frequently disagree on how to identify each fallacy. This poses an identification problem that is prior to questions of analysis and evaluation of the fallacies.

Before going on to propose a theory of fallacy, a general analysis of the concept of fallacy, it is necessary to make some comments on these three problems. These three problems cannot be fully solved until we have a theory of what a fallacy is generally. But before we can even intelligently discuss or put forward such a theory, we must outline some basic considerations with respect to these three problems that need to be taken into account.

1. Argumentation Schemes and Themes

The formal and inductive fallacies outlined in chapter 3 can be evaluated with respect to deductive and inductive forms of reasoning that are currently familiar to logicians. Of the informal fallacies outlined in chapter 2, eleven clearly need to be evaluated in relation to the argumentation schemes for presumptive reasoning presented in chapter 5.

1. Ad Misericordiam
2. Ad Populum
3. Ad Hominem
4. Straw Man (Argument from Commitment)
5. Slippery Slope (all four types)
6. Argument from Consequences
7. Ad Ignorantiam
8. Ad Verecundiam
9. Post Hoc
10. Composition and Division
11. False Analogy

But there are eight informal fallacies remaining that do not fit any of the argumentation schemes.

1. Equivocation
2. Amphiboly
3. Accent
4. Begging the Question
5. Many Questions

6. Ad Baculum
7. Ignoratio Elenchi (Irrelevance)
8. Secundum Quid (Neglecting Qualifications)

The fallacy of secundum quid can occur with any kind of presumptive argumentation. Presumptive reasoning is inherently defeasible and subject to exceptions. Where proper qualifications are ignored or suppressed, and this can happen with any kind of presumptive argumentation, the fallacy of neglecting qualifications occurs.[1]

Irrelevance and begging the question are fallacies that have to do with sequences of extended argumentation where the links of inference making up the reasoning can be of different kinds, including any types represented by the various argumentation schemes. And equivocation (including amphiboly and accent) can occur with respect to any kind of argumentation. Hence none of these eight fallacies is tied to any particular argumentation scheme.[2]

The evaluation of these fallacies requires studying the use of an argument in a broad context of dialogue. At this level, we need to ask within what type of dialogue the argument is supposed to be taking place and in what stage of the dialogue it is. This is called the *dialectical* level of analysis, because it pertains to how the argumentation was used in a context of dialogue to contribute to the goal of that type of dialogue. For these eight fallacies especially, this level of analysis is crucial. But with all the major informal fallacies studied in this book, it will be seen that this dialectical level of analysis is necessary to some extent.

Typically, in order to evaluate an argument as correct or fallacious, in addition to seeing whether it meets the requirements of its appropriate argumentation scheme, we also need to see how it is used over a larger segment of the dialogue, in the sequence of question-reply argumentation. To do this, we need to reconstruct a profile of dialogue in which the argumentation should properly be used and to contrast this with the actual sequence that took place. This task involves the application of a normative model of dialogue to a test of discourse given in the actual case to be analyzed (as far as it can be reconstructed). Good examples are the profiles of dialogue used to analyze the fallacy of many questions in Walton (1989b) and the profiles of dialogue used to analyze the fallacy of begging the question in Walton (1991a).

As contrasted with an argumentation scheme, which is a local inference used at one point or stage of a dialogue, an *argumentation theme* is a sequence of argumentation modeled in a profile of dialogue that reveals how the argument was used in a protracted manner over an extended stretch of dialogue (longer than a single scheme but generally shorter than a complete dialogue from the opening to the closing stage). Usually an argumentation theme will be a subsequence of the pairs of moves in the argumentation stage.

For each argumentation scheme, there is a matching set of critical questions appropriate for that scheme. To ask an appropriate critical question in a dialogue shifts the burden of proof back onto the side of the proponent of the original argument to reply to this question successfully. For each use of an argumentation scheme by a proponent of an argument in a dialogue, typically there arises a whole sequence of questions and replies that arise from the response of the respondent and the subsequent replies of the proponent. This sequence of connected arguments, questions and replies, is called the argumentation theme.

By studying the argumentation theme in a given case, in relation to the requirements of a normative model of dialogue appropriate for that case, we can learn much about the critical attitudes of the proponent and the respondent. For example, we can ask whether the proponent is putting forward her argumentation in a way that shows that she is observing the Gricean maxims of honesty, cooperativeness, relevance, and so forth for that type of dialogue, like a critical discussion, or whether she is not really open to paying due accord to the evidence put forward by the other side but is merely engaging in eristic dialogue or quarreling. Such a judgment is generally best made not at too localized a level, on the basis of a single inference or putting forward of an argumentation scheme, but rather on the basis of performance over a longer, protracted sequence of dialogue exchanges.

2. The Fallacy of Many Questions

The classic case of the argumentation theme fallacy is the fallacy of many questions. When we ask a question with presuppositions like "Have you stopped cheating on your income tax?" (case 17), we presume that a sequence of argumentation moves in the prior sequence of dialogue has already been set into place in a certain type of profile. First, it is assumed that the questioner has asked the respondent whether she has made income tax returns in the past. Then it is assumed that the questioner has asked the respondent whether she has cheated on those income tax returns in the past. Only then is the questioner justified in asking the respondent, "Have you stopped cheating on your income tax?"

The profile of dialogue that represents the normatively correct sequence of dialogue for this question to be asked is shown in figure 3. Then the profile of dialogue could also follow the answering of the

	QUESTIONER	RESPONDENT
1.	Have you made income tax returns in the past?	Yes.
2.	Have you cheated on those income tax returns in the past?	Yes, I admit it.
3.	Have you stopped cheating on your income tax?	

Figure 3   Sequence of Concessions to the Tax Question

key question, one way or the other.[3] For example, if the respondent answers "Yes" at round 3, then at question 4 she could be asked, "Do you know that cheating on your income tax is a crime?" Here we have a sequence of adjacency pairs of question-reply argumentation moves in a dialogue that are tied in together in a connected argumentation theme. The tableau, or sequence of moves represented in figure 3, is called a profile of dialogue.

A profile of dialogue is something less than a whole dialogue but something more than a single move, like a single question, reply, or argument in a dialogue. It represents a connected sequence of moves, an argumentation theme that shows how the sequence of argumentation should go in a dialogue if it is to be correct and nonfallacious. Mixing up these moves in a certain characteristic way can be identified with a type of fallacy.

The fallacy of many questions (complex question) is a kind of tactic designed to entrap a respondent by asking him a complex question that has propositions built into the questions as presuppositions that are very damaging to the respondent's side of the dialogue and where, as soon as he gives any direct answer to the question, he becomes committed to these propositions.[4] The fallacy is a violation of asking the sequence of questions in the right order, as modeled by the profile of dialogue for asking the given question. The fallacy is a failure to secure commitment first to the propositions presupposed in the complex question. It is a fallacy of asking too much at once. By preventing the respondent from answering fairly (without questioning the question), the respondent is inhibiting the right kind of dialogue needed to elicit the respondent's commitments and thereby have an exchange that contributes to the goal of the dialogue. This is the real basis upon which we should call the fallacy of many questions a fallacy.

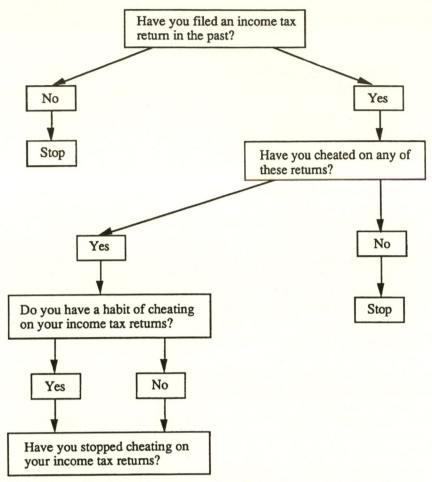

Figure 4 Profile of Initial Sequence of Dialogue

The profile of the initial sequence of questions and replies exhibited in the flow chart in figure 4 functions as a partial normative model by showing the correct type of sequence of argumentation the relevant part of the dialogue should take. This sequence must, however, be placed in a larger context of dialogue before we can fully evaluate whether the given question (as used in context, in a particular instance) is fallacious or not.

The profile of subsequent sequence of dialogue for the same question shows, as given in figure 5, how this question can be used as a sophistical tactic to get the best of a partner in dialogue by getting him to make a concession that counts heavily against his side in the dialogue exchange. A fuller analysis of the fallacy of many questions

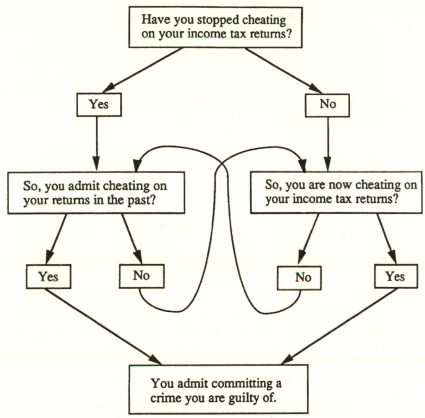

Figure 5   Profile of Subsequent Sequence of Dialogue

is given in Walton (1989b, chap. 2). But enough evidence has been given here to show how this fallacy needs a thematic kind of analysis.

3. Begging the Question

Begging the question is another fallacy that is not to be identified with the abuse of any particular argumentation scheme or valid form of argument. Indeed, the form of argument most characteristic of arguing in a circle—see case 20—is 'A, therefore A,' which is deductively valid. Begging the question is a fallacy because there is a sequence of argumentation, a sequence of questions and replies, that goes in a circle, as shown by the profile of dialogue. In the case of this fallacy, the most useful way of tracking the circular sequence of reasoning in the dialogue is the technique of argument diagramming.

BANK MANAGER	SMITH
1. Can you give me a credit reference?	My friend Jones will vouch for me.
2. How do we know *he* (Jones) can be trusted?	I assure you he can.
3. Yes, but how do we know *you* can be trusted?	[Back to reply at round 1.]

Figure 6   Question-Reply Sequence in the Bank Manager Case

But standing behind the argument diagram is a profile of dialogue modeling the argumentation theme in the given case.

For example, in case 21, there is a sequence of questions and replies, represented in the profile in figure 6. We can see in figure 7 how the argumentation theme is essentially circular. It keeps looping back to the same point. Its argument diagram has essentially the structure shown in figure 7. There is an argumentation scheme involved in this case—it is a form of argumentation from testimony where one person, who is presumably a reliable person, is asked to vouch for the reliability of another person whose reliability is in question. But the fallacy of begging the question does not rely on any single failure of this argumentation scheme to be presumptively reasonable, by itself. The fallacy comes in when you put the chain of argumentation schemes, as used in case 21, together in a sequence of connected dialogue. The resulting circle, shown in figure 7, when the

①: Smith can be trusted.

②: Jones vouches for him.

③: Jones can be trusted.

④: Smith vouches for him.

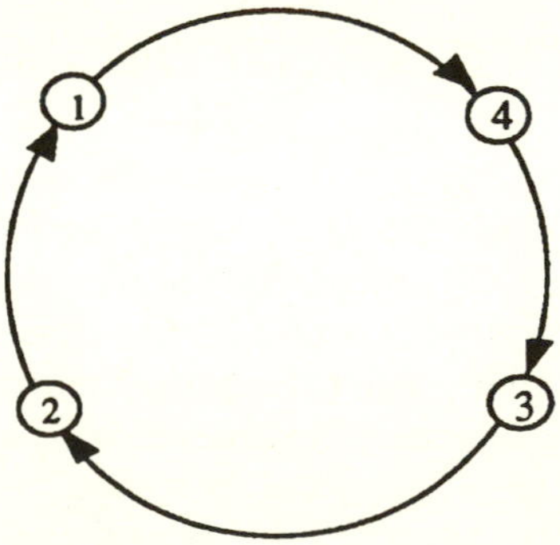

Figure 7   Graph of Reasoning for the Bank Manager Case

schemes are joined together in an argumentation theme, indicates the fallacy. No independent reason is given why either party should be trusted, without relying on the trustworthiness of the other party (which is in question). If you look at the sequence of argumentation as a whole, it establishes nothing, because what was in question is still in question at the end of it all.

A comparable kind of analysis applies to the sequence of question-reply dialogue in case 22. Here the sequence of questions and answers takes us in a circle too. Here the profile of dialogue can be represented by figure 8 below. Here an argumentation scheme is also involved. The believer is invoking argumentation from authority of something like expert opinion, which could perhaps be called a kind of argumentation sacrosanct testimony of some sort. It is not the correctness or incorrectness of the use of this single move in the argument, however, that makes case 22 an example of the fallacy of begging the question. It is how the whole sequence of questions and replies is connected together as an argumentation theme.[5] The profile of argumentation shows how the longer sequence of questions and replies has gone in a circle.

Once again, it is not the circular reasoning in the case that is fallacious per se. The circular reasoning could be nonfallacious, for example, if the nonbeliever is not a total skeptic about religion but a person whose faith is a little shaky, who basically accepts the Bible as a divine source that is true and reliable but who doubts whether this source definitely says or proves the existence of God. See Colwell (1989) for an elaboration of this possible interpretation.

What makes the circular reasoning in this case fallacious is essentially the same failure of evidential priority in the argumentation theme as found in case 21. To prove that God exists to someone who is thoroughly skeptical about this proposition, and the whole reli-

BELIEVER	NONBELIEVER
1. God exists.	How do you know?
2. The Bible says so.	How do I know what the Bible says is true?
3. The Bible is the word of God.	How can I accept that without already presuming that God exists?

Figure 8 Question-Reply Sequence for the God and the Bible Case

gious point of view based on it, you can't present a chain of argumentation that comes back to a reliance on the acceptance of the existence of God as an authoritative source.

Of course, in cases 21 and 22 the outrageously fallacious use of circular argumentation is pretty obvious. These are cases meant to illustrate the fallacy for textbook purposes; they are not cases where digging out the fallacy is meant to be difficult. In the more difficult cases—arguments that might actually deceive someone in a serious dispute—the chain of argumentation is much longer and more complicated, so that the circular argumentation theme is concealed.[6] In such a case, for example, a circular argument could be spread over a chain of reasoning in a lengthy article or even a whole book.

According to the analysis of the fallacy of begging the question given in Walton (1991a), there are three stages in evaluating a given argument to judge whether this fallacy has been committed or not. First, you have to reconstruct the sequence of reasoning in the argument by making an argument diagram of the premises and conclusions in the reasoning. This step shows whether there is a circle in the argument diagram and exhibits the structure of the circular reasoning. But not all circular reasoning is fallacious.

The second step is to determine the context of dialogue and the purpose of the argument as a contribution to a dialogue. This is important, because circular argumentation is always excluded in an inquiry, for example, but could be much more acceptable in a critical discussion.

The third step is to determine, given the context, whether the argument is meant to have a *probative function,* meaning that the premises are supposed to be used as an evidentiary basis for proving, or building support for, a conclusion that is (at least initially) more doubtful than these premises. An argument begs the question if it fails to fulfill a probative function because of its circularity, as used in a context of dialogue where it was supposed to fulfill such a probative function. At any rate, this should be enough to convince the reader that begging the question is a thematic fallacy.

Enough said about thematic fallacies for the moment. Let us begin a consideration of those fallacies that relate to specific argumentation schemes as well. Here we confront the problem of the names of the fallacies.

4. Fallacy Names

Some of the types of arguments described by the names given to the fallacies in the traditional textbook treatment are always fallacious, and some of them are kinds of argument that could be either reasonable or fallacious. With some of them, different terms are used by different textbooks. Whether the argument falls into one category or the other depends on what term you use.

For example, begging the question (petitio principii) is always fallacious. An argument that begs the question is a fallacious argument. Circular reasoning is not always fallacious, however. Sometimes it is not fallacious (Walton 1985b; 1991a). A circular argument is fallacious when it begs the question in a given case.

Argumentum ad verecundiam literally translated means argument to respect (reverence). So described, it would appear to be a fallacy.[7] The description 'appeal to expert opinion in argument' refers to a kind of argumentation that can be reasonable in many instances. Appealing to expert opinion is a legitimate presumptive type of argumentation. It can go wrong, or be used wrongly, of course, in some cases. And the phrase 'argument to reverence' suggests exactly the type of case where it so often is used wrongly, namely one where one party tries to take advantage of the submissiveness of the other party to some supposed expert or authority the other party is in awe of. This suggests a kind of tactic of abuse of appeal to authority that is generally fallacious.[8] What is very confusing here is that many of the textbooks simply equate argumentum ad verecundiam with "appeal to authority" or "appeal to expert opinion" and translate the Latin expression using one of these phrases. You could say the same thing about many of the other fallacies.

Some of the classifications in figure 9 are debatable, depending on how you translate or analyze the phrase used as the traditional name of the fallacy. *Ad misericordiam* translated as "appeal to pity" sounds somewhat illicit, because pity is often felt to be a condescending attitude or emotion. If you rechristen it "appeal to sympathy," however, it begins to sound much more acceptable.[9] Even so, appeal to pity, despite its somewhat negative connotations, is a commonly accepted type of argumentation and can be quite reasonable in its proper place.

Post hoc ergo propter hoc is another one that is a little tricky. If you take post hoc as referring to arguing from correlation to causation, then this type of argumentation is, in many instances, quite legitimate and nonfallacious. It is better (Walton 1989a, 212–33), however, to think of post hoc as the fallacy of leaping too quickly to a causal conclusion on the basis of a correlation without taking other relevant factors into account. So conceived, post hoc belongs in the 'fallacious' column.

Many questions belongs in the other category, because asking a question like "Have you stopped cheating on your income taxes?" could be reasonable in the right circumstances, for example, if the

CAN BE REASONABLE OR FALLACIOUS	FALLACIOUS
Ad Hominem	*Ad Verecundiam*
Slippery Slope	Equivocation
Ad Ignorantiam	Amphiboly
Ad Baculum	Accent
Ad Misericordiam	Begging the Question
Many Questions	*Ignoratio Elenchi*
Ad Populum	Straw Man
Argument from Consequences	Hasty Generalization
Composition and Division	*Post Hoc*
	False Analogy

Figure 9   Division of Traditional Fallacies

respondent just conceded cheating on his income tax returns while being cross-examined in a legal trial situation (as shown in section 2 above).

Ad populum can be quite a reasonable type of argumentation if you translate it as "appeal to popular opinion." Some texts use pejorative phrases like "mob appeal" or "appeal to the gallery," however, to characterize this fallacy.[10] These phrases strongly suggest a type of argumentation that is inherently fallacious. All *ad populum* means, however, is "to the people," suggesting a kind of argumentation that could be nonfallacious in many cases, at least as a form of presumptive reasoning, subject to default.

The phrase *secundum quid* means "in a certain respect," which could be taken to describe presumptive reasoning generally, a kind of reasoning that is inherently subject to qualifications because of its nonabsolute or nonuniversal nature.[11] If you translate this phrase as "neglect of qualifications," however, or use that phrase to name the fallacy, then the type of argumentation described would seem to be generally fallacious. Also the common names "hasty generalization," "overgeneralization," and so forth, strongly suggest an inherently fallacious type of argumentation.

The problem of agreeing on a standard name is particularly acute with this fallacy. Instead of the more descriptive terms 'hasty generalization' or 'neglecting qualifications,' many textbooks still use antiquated and misleading terms like 'accident,' 'converse accident,' and so forth.

The names of the linguistic fallacies like equivocation and amphiboly definitely stand for "fallacious." Presumably, an equivocal argument, or one containing an equivocation, must be a fallacious argument. Indeed, an argument containing an equivocation is really not one argument at all but several that only appear to be one because of an ambiguity in the language in which they have been put forward.[12] This multiplicity already suggests a deception or confusion that goes against the goals of a properly run dialogue where one argument is put forward at a time. Hence equivocation belongs in the 'always fallacious' category, and the same could be said for amphiboly and accent.

A straw man argument is always fallacious, because any distortion or misrepresentation of another party's position is a bad thing in argumentation. In a critical discussion, for example, if one party has a distorted or incorrect representation of the other party's point of view, this could be a strong impediment to resolving their conflict of opinions.[13] Perhaps even more crucially, it would prevent proper maieutic insight into one's own point of view from developing. For such insight, presumably, requires a contrast to develop between the two conflicting points of view, a contrast that reflects the real differences between the two points of view.

Ad baculum arguments are generally fallacious in a critical discussion but not in other types of dialogue. The negotiation dialogue is a case in point. From the point of view of van Eemeren and Grootendorst (1992), however, ad baculum arguments in the form of threats or appeals to force may always be regarded as fallacies. Clearly threats and appeals to force have no place in a critical discussion, and their only function is to block the goal of dialogue by preventing appropriate argumentation from being put forward. Threats, however, especially indirect threats to impose sanctions, are an accepted part of the bargaining tools in many negotiations.[14] In such a context, an argumentum ad baculum is not necessarily fallacious.

For this reason we categorize ad baculum arguments under the heading of not always fallacious. If you consider only the critical discussion, like van Eemeren and Grootendorst, however, then you would put them in the other column. Evidence presented in section 7 below will show, however, that this approach is not fully adequate to analyze ad baculum as a fallacy.

5. Classifying *ad Hominem* Arguments

There is no definite agreement in the textbooks on how to classify the various types of ad hominem arguments. Some call the bias type of ad hominem "circumstantial," while others call the circumstantial type *tu quoque*, and so forth. Some classify the bias type as a subtype of the circumstantial.[15] Now, however, on the basis of the argumentation schemes set out in chapter 5, a basis for classification can be given as follows.

There are three basic types of ad hominem argumentation and two derived subtypes that also bear discussion in this category. One is the direct or abusive ad hominem argument, which is essentially negative argumentation from ethos, used to criticize someone's argument by claiming the person has bad character. This type of argumentation is an attack on a person's sincerity as a cooperative participant in a dialogue. Another is the circumstantial ad hominem argument, which essentially corresponds to the argumentation scheme for the circumstantial argument against the person in chapter 5. This type of argument is a species of argumentation from commitment that involves a clash (or practical inconsistency) between what a person claims in his argument and what he is committed to according to his known personal circumstances. The point of essential difference between the abusive and circumstantial arguments is that the latter, but not the former, requires an alleged clash (or practical inconsistency) of commitments. Of course, in practice, very often the two are connected, and typically, for example, the circumstantial ad hominem is a lead-in to the abusive ad hominem. A typical case would be the following.

> *Case 104*
>
> Well, Smith always says that the opposition party misuses government funds to live an opulent lifestyle, and he is saying how horrible that is. But I happen to know that Smith himself went on a skiing vacation paid for by government funds. That man Smith is a hypocrite. You can't believe a word he says.

Here the circumstantial ad hominem leads into an abusive ad hominem. This transition is common in ad hominem argumentation, but we can still distinguish the circumstantial and direct (abusive) as two distinct types of argumentation.

The third type of ad hominem argument is the bias type. Many textbooks classify this as a species of circumstantial ad hominem argument. We can now see, however, that the bias type is essentially different from the circumstantial. The bias type is an instance of the argumentation scheme for the argument from bias (chapter 5, section

8). The bias type of ad hominem argument is not a species of argumentation from commitment, and it does not essentially involve or require a conflict of commitments alleged on the part of the person attacked. Instead bias, as defined by Walton (1991b), is a failure of critical balance in a dialogue, such as a critical discussion, that requires a participant to take into account the argumentation on both sides of an issue and be willing to concede to a good argument that supports the opposed side.

Another type of ad hominem argument recognized by many textbooks is the tu quoque. This is not really a separate and distinctive type of ad hominem argument in its own right. It is rightly emphasized by the textbooks as a phenomenon, because one of the worst dangers of using any ad hominem argument is that your opponent will reply in kind, leading to a quarrel. For example, in case 4 a woman argues that a man is biased, simply because he is male and therefore can never see an issue like abortion from the woman's point of view. But the person so criticized, a man, could turn the argument on its head (tu quoque) and argue that women can only see it from the women's point of view and are therefore biased themselves. Here the danger is a standoff or deadlock where both sides have been disqualified from continuing a critical discussion of the subject. The resulting polarization leaves only quarreling open as a way of continuing the dialogue. And that is most often the outcome of such a tu quoque move.

The tu quoque reply is common with many kinds of argumentation, however. For example, in the ad ignorantiam, we can have participants arguing: "You prove it! You disprove it!" Or in the ad verecundiam type of argumentation, we can have exchanges like: "You are not an expert! Well, you're not either, so there!" and so forth.

This leads us to the fourth type, the poisoning-the-well type of ad hominem. This type of ad hominem argumentation is best seen as an extension of the bias type of ad hominem, except that the bias is alleged to be of a type that the person attacked can never change. For example, in case 4, the person attacked is said to be biased because he is a man. This gender bias is something he cannot change (at least, practically speaking, with respect to the argument at issue). Therefore, no matter how fair he is in argumentation, or how open he is ready to be to opposing views, still whatever he says in the future is always apparently clouded by this bias that he can never get rid of or transcend. This is a kind of muzzling attack that disqualifies a person from taking part in a critical discussion at all, with any credibility, hence the appropriateness of the phrase "poisoning the well." The poisoning-the-well type of ad hominem could also be seen as an extension of the abusive or the circumstantial types of ad hominem in some cases. Most typically and characteristically, however, it is an extension, or extreme form, of the bias type of ad hominem argument.

Note that the basis for distinguishing the three basic subtypes of ad hominem is not that they are four distinctive fallacies because they violate four distinctive rules for a critical discussion or any other type of dialogue. Instead, the first three are classified as distinctive subtypes of ad hominem argumentation because each of them has a distinctive argumentation scheme. Each of the three is a kind of argumentation that can be presumptively correct, or can be used correctly, in a given case. It is this positive aspect that is the right basis for distinguishing each of them as a main subtype of ad hominem argumentation.

The tu quoque then comes out as a type of tactic or profile of dialogue often associated with the use of the ad hominem argument and worth noting as a danger connected with it. It is a way of extending any of these three types of ad hominem argumentation further in a profile of dialogue. But it is not a type of profile that is exclusive to the ad hominem argument or essential to it as a distinctive kind of argumentation.

The poisoning-the-well variant comes out as a kind of extension of the other three types of ad hominem arguments (but especially, and characteristically, of the bias type). Although it does not have a distinctive argumentation scheme among the schemes analyzed in chapter 5 (and rightly so, because it is not, in itself, an acceptable kind of argumentation except in the quarrel and certainly not in the critical discussion), it can be seen as a tactic built on and exploiting the use of the argument from bias.

The poisoning-the-well type of ad hominem argument violates rule 1 of van Eemeren and Grootendorst (1992, 208), which forbids parties from preventing each other from advancing or casting doubt on standpoints. But that does not define it as a fallacy, for it fails to specify the tactic used to prevent a party from advancing or casting doubt on a standpoint in a dialogue. The tactic used is to argue that the other party in such a dialogue is so consistently or hopelessly biased that she always argues only from her own closed point of view or according to her own special interest. This type of attack amounts to a claim that the other party is really engaged in a quarrel or in interest-based bargaining type of dialogue when she is supposed to be, or purports to be, engaged in a critical discussion or in another type of dialogue where too strong a bias toward one's own side blocks the goals of the dialogue.

The real basis of the poisoning-the-well type of ad hominem argument is therefore, in light of chapter 4, best analyzed as a dialectical shift between two types of dialogue. This also is true of the argument from bias generally. For example, in case 3, the basis of the bias ad hominem argument is that Smith is really (covertly) engaged in interest-based negotiation or bargaining on behalf of business rather than in a critical discussion on whether higher taxes will contribute to the recession. The allegation is that he is not really open to taking the arguments on both sides into account.

6. The *ad Hominem* Fallacy

As a fallacy, the ad hominem argument relates to failures of its use to respond appropriately to critical questions for three argumentation schemes—the negative ethotic argument, the circumstantial argument against the person, and the argument from bias. But not every such failure is a fallacy.

There are three main ways an ad hominem argument tends to fail and thereby constitute a fallacy. One is that the premise (alleging bad character, practical inconsistency, or bias) may fail to be sufficiently backed up by evidence, and the strength of the ad hominem may be greatly exaggerated (for example, by innuendo). The second is the basic ad hominem fallacy, analyzed in Walton (1989a, chap. 6), as a shift from weak refutation to strong refutation. The third is the failure of the attack to be relevant (in context of dialogue).

Hence, if we describe the ad hominem fallacy as rule violation in a critical discussion, it is, or could be, a violation of any of three rules. The first failure is a failure to fulfill burden of proof. The second failure is what van Eemeren and Grootendorst (1987, 291) call absolutizing the failure of a defense. And the third is a violation of the rule of relevance. Thus the ad hominem fallacy does not correspond to the violation of any single rule of a critical discussion. What is also interesting is that all three types of ad hominem argument, the direct, the circumstantial and the bias type, prominently exhibit the same three types of failures of correct argumentation.

Another interesting thing to note here is the difference between failures of this kind and fallacies. For example, a failure to back up an allegation of bad character is not, in itself, a fallacy or a fallacious argument. It is a failure to meet burden of proof, and therefore it is a violation of a rule of a critical discussion but not necessarily a fallacy or a fallacious argument (for this reason alone). There are two other requirements. One is that the personal attack on character must be used to argue that the person's argument is wrong. The attack on character must be used to run down the person's argument—that is, the argument of the person whose character has been attacked. The other requirement that has to be met is that the failure to meet burden of proof has to be a more serious failure than just an insufficiently supported premise. For that could be just a blunder, as opposed to a fallacy, or just a weak argument in the sense of being open to criticism. And a weak argument, in this sense, is not necessarily a fallacious argument. For to say that an argument is fallacious is a strong form of condemnation.

Typically, an ad hominem of the direct (or abusive) type is fallacious not just because the allegation of bad character is insufficiently supported by the attacker. The reason is that this type of argumentation from negative ethos is put forward as an innuendo or suggestion where the attacker leaves open plausible deniability. The aim is to raise a cloud of suspicion against the person attacked while evading any need to have to give evidence to back up the charge. The charge can be a vague one like "wallowing around in the sewers so long, he doesn't know the difference between right and wrong," that was never meant to be backed up by specific hard evidence. Or it can be put forth on the basis of innuendo like, "Well, I heard someone the other night say that Smith cheats on his tax returns, but I can't say who said that, and of course I don't believe it myself." This type of attack is an ad hominem fallacy, as opposed to simply a blunder, or a weak argument from negative ethos, because there is a systematic tactic used by the attacker to shield himself from ever having to fulfill burden of proof. Yet at the same time the damage is done by raising a cloud of suspicion.

Another requirement of our analysis of the ad hominem fallacy is that it must be an argument. In other words, a simple personal attack, or an allegation of bias or circumstantial inconsistency, should not be classified as an ad hominem fallacy, even if the attack is unwarranted, unfair, immoral, illegal, or otherwise open to condemnation. For it to be an ad hominem fallacy, it must be an ad hominem argument.[16] That is, it must be an instance of one of the argumentation schemes, of the three outlined in chapter 5, that identifies it as a species of ad hominem argument.

The general issue posed here is whether a fallacy must be a fallacious argument or whether it can be any type of move or speech act that violates a rule or maxim of politeness in a dialogue. Ultimately it will be made a general requirement of the analysis of the concept of fallacy advocated in this book that a fallacy must be a fallacious argument (or at any rate, something that is supposed to be an argument).

A sharp contrast in the way of analyzing a fallacy can be illustrated in the way the ad hominem fallacy is dealt with in our analysis of it and that of van Eemeren and Grootendorst (1992, 212). In the latter analysis three types of ad hominem fallacy are described below.

1. Argumentum ad hominem (direct personal attack, abusive): Doubting the expertise, intelligence, or good faith of the other party.
2. Argumentum ad hominem (indirect personal attack, circumstantial): Casting suspicion on the other party's moves.
3. Argumentum ad hominem (tu quoque): Pointing out an inconsistency between the other party's ideas and deeds in past and/or present.

Each of these three is said to be a fallacy that occurs at stage 1 of a critical discussion (the opening stage). And each of these three is said to be a fallacy because it violates rule 1: "Parties must not prevent each other from advancing standpoints or casting doubt on standpoints." We should note also that ad misericordiam (213) and ad baculum (212) are said to be fallacies by van Eemeren and Grootendorst on the grounds that they violate rule 1 at stage 1 of a critical discussion.

This analysis is quite different from mine. On my analysis the argumentum ad hominem is a kind of argumentation that often, even typically, occurs at the argumentation stage of a dialogue rather than at the opening stage. In my analysis, the ad hominem is in fact in many cases a legitimate type of argumentation in its own right. It is both a common and a powerful type of argumentation in its own right (as indicated by its argumentation schemes for its subtypes) and not just an opening move in a dialogue.

On my analysis, the essence of the ad hominem is not captured by saying that it prevents someone from advancing a standpoint or casting doubt on a standpoint. Sometimes it has this effect, but so do many of the other fallacies, in many cases, have this same effect. On my analysis, there are four types of ad hominem argument, and our way of defining and classifying them is quite different from the categorization given by van Eemeren and Grootendorst.

We define the abusive or direct personal attack type of ad hominem as the rejection of someone's argument on the grounds of his bad character, especially his character for veracity. We do not see this type of argument as fallacious per se. It is a negative argument from ethos that is, in principle, a legitimate kind of presumptive argumentation with its own characteristic argumentation scheme. It can be abused, or used fallaciously, in various ways. But two are prominent. One is that the premise that the party attacked has bad character may not be supported by adequate evidence. And indeed, it may be false or may even be used as an unfair innuendo attack with no real basis either forthcoming or intended. The other way it is used as a fallacy is failure of relevance. Typically in this kind of case, the ad hominem is a quarreling type of argumentation that would be appropriate in a quarrel but is used to block a critical discussion. If the dialogue in question is supposed to be a critical discussion, for example, then quarreling may be an exciting distraction that upsets everyone but distracts from the real purpose of the discussion.

On my view, the personal ad hominem is not the same as doubting the expertise or intelligence of the other party. It is attacking the character of the other party and using that as a basis for saying you can't trust him to follow the principle of cooperativeness needed to take part in a dialogue like a critical discussion as opposed to a quarrel.

What they call the tu quoque is what I call the circumstantial ad hominem argument. What they call the "indirect" or "circumstantial" (casting doubt on motives) does not correspond exactly to any of my categories but might fall somewhere near or in the area of what I call the bias ad hominem argument.

Again, with both subtypes of ad hominem, I do not see them as fallacies per se, nor do I see them as fallacious because they violate a single rule of dialogue. I see them as arguments that can be used fallaciously in various ways. But I see them as ad hominem fallacies because they each represent an argumentation scheme that can be used wrongly or inappropriately in a context of dialogue. So used, they can actually impede (or even block) the progress of the dialogue instead of contributing to the achievement of its goal.

7. The *ad Baculum* Fallacy

The first problem with the ad baculum fallacy is defining it as a type of argumentation. Some textbooks define it as the use of a threat to cause acceptance of a conclusion. Others define it as the appeal to force in argument (Rescher 1964, 79, and Cederblom and Paulsen, 1982, 100). Still others (Engel 1976, 130) define it as the fallacy of appeal to fear. Some textbooks combine two of these categories. Damer (1980, 91) calls the fallacy "appeal to force or threat." Others divide the categories into separate fallacies. Michalos (1970) has appeal to force (50) as one fallacy and appeal to fear (*argumentum ad metum*) as a separate fallacy (58).

The variety of treatments leaves open various ways of defining the ad baculum argument. Wreen (1988) thinks that a threat is not essential to ad baculum and that the argument might work by appealing to fear without a threat, as in intimidation or scaremongering tactics. The scheme of classification given in Walton (1992b, 180) agrees with this approach, distinguishing between a narrower type of ad baculum argument that essentially involves a threat and a wider type that does not but still appeals to fear (intimidation).

Another problem with the ad baculum fallacy is to determine how it involves, or is based on, argument. For a threat, or appeal to fear, is not necessarily an argument. But the most common type of ad baculum fallacy, and probably the most effective and deceptive, is the indirect type where what is overtly a warning functions covertly as a threat. This type of argument is a species of argumentation from consequences taking the form: "If you do (or fail to do) some action, the bad consequences (bad, from your point of view) will follow." A typical example is case 6, where the speech act would be clearly evident to the respondents as a threat, even though, on the surface, the speech is put in the form of a warning. According to the speech act analyses of threat and warning given in Walton (1992a, 169–74), the argument in this case is a threat because the speaker is indicating his willingness to the respondent to carry out the "unfortunate consequences" if the respondent does not act in the way proposed.

An argument using a threat is so clearly inappropriate in a critical discussion that, in principle, it can be immediately classified as a fallacy. For by its nature, the argumentation in a critical discussion must have an open quality, so that any argument can be subjected to critical questioning and freely objected to or rejected if there is not sufficient evidence given to support it. In this context, a threat used to try to gain acceptance of a conclusion is highly inappropriate. One might ask, therefore, why the ad baculum is such an effective tactic that is commonly and plausibly used to deceive arguers as a fallacy. To grasp why, one must look carefully at the context of dialogue.

In negotiation dialogue, the threat of action or sanctions is recognized as a legitimate bargaining tactic. It is normal in negotiation to put forward argumentation of the form, "If you give me such-and-such as a concession, then I'll concede this other thing to you in return." The doubly negated form of this conditional is also quite common and generally acceptable as a form of argument in negotiation: "If you don't do such-and-such, then I won't do such-and-such other thing, and that will be very bad, from your point of view (a loss)." Of course, overt threats of this type may be perceived as inappropriate, impolite, or intolerable. But generally such threats are put forward in an indirect (covert) way, as a species of argumentation from consequences, that masks or softens their hard edges. As Donohue (1981a, 279) puts it, threats are high risk tactics in negotiation dialogue because they are perceived as a final stand, but they are core concepts of bargaining theory and are, in themselves, legitimate as arguments in negotiation. In short, threats used in arguments are not fallacious per se in negotiation dialogue, as contrasted with persuasion dialogue, where use of a threat may generally be presumed to be fallacious. It could be added as well that threats are common and generally nonfallacious in eristic dialogue, especially in the quarrel.

It is possible to have an argumentation scheme for the use of a threat in argumentation. It is a species of argument from consequences used as a kind of practical reasoning to try to get a respondent to carry out (or stop from engaging in) a certain action. The major premise takes the form, 'If you (respondent) do (or fail to do) A, then something B will follow as a consequence, and B will be bad, from your point of view.' Such a type of argumentation really has no legitimate place in a critical discussion, however. It is only valid, or used appropriately, in other types of dialogue like the negotiation or the quarrel.

Hence for the ad baculum fallacy, at least for the central or narrower type that essentially involves a threat, the key factor in evaluating a case as fallacious or nonfallacious is the context of dialogue. The very same appeal to a threat in an argument could be fallacious as used in one context of dialogue, like a critical discussion, but nonfallacious as used in another context of dialogue, like a negotiation.

Indeed, it is the dialectical shift from one context of dialogue to another that is the basis for explaining the ad baculum as a fallacy. In particular, it is this type of dialectical shift, especially when it is concealed and/or illegitimate, that solves the puzzling problem of why the ad baculum is such a common and effective tactic as a sophistical deception. Of course the ad baculum argument is outrageously inappropriate in a critical discussion, but if it is unclear whether the dialogue is really supposed to be a critical discussion or a negotiation, or if there has been a shift from the one type of dialogue to the other, the inappropriateness of the argument is masked or confused, giving it a semblance of correctness or legitimacy.

Hence once one has identified an ad baculum argument correctly by identifying the threat, very often expressed indirectly using argumentation from consequences, the key next step is to ask the question, "What type of dialogue were the participants originally supposed to have been engaged in?" From there to achieve an evaluation of the argument as fallacious or not, one needs to examine the textual evidence of the existence of a dialectical shift. In some cases, it is quite a subtle question to determine whether the ad baculum argument can correctly be judged fallacious or not. For example, in political argumentation, typically negotiation dialogue is involved, even though, at some normative level of ideality, the dialogue is supposed to have elements of a critical discussion.

In short, according to the analysis laid out here, the ad baculum argument should not always immediately be presumed to be a fallacy. But when it is a fallacy, it is a dialectical failure, an inappropriate use of argumentation in relation to the type of dialogue in which the participants in the talk exchange are supposed to be engaged. In other words, it isn't the argument itself, as an argumentation scheme, that is fallacious or not. It is a matter of how that argumentation has been used in a context of dialogue. That turns out to be a basic characteristic common to all four of the *ad* fallacies studied in this chapter, strongly suggesting that matters of analyzing the argumentation theme in a given case are very important for all four.

8. The *ad Misericordiam* Fallacy

Before attempting to evaluate how and why the argumentum ad misericordiam is a fallacy, there exists a definition or translation problem of determining exactly what is supposed to be meant by *misericordia*. The *Oxford Latin Dictionary* (Glare 1982, 1118) gives the following definition: "misericordia ~ae, f. [MISERICORS + -IA]: **1.** Tender-heartedness, pity, compassion. **2.** Appeal to compassion, pathos." On the question of whether an argument that appeals to or depends on *misericordia* is fallacious or not, much depends on exactly how this word is translated into English.

The usual nomenclature is to translate *argumentum ad misericordiam* as "appeal to pity," but the word 'pity' has negative connotations. For example, if you ask a disabled person whether he wants your pity, he is likely to reply negatively because 'pity' implies he is somehow an unfortunate or lesser person. He would be more likely, however, to welcome your sympathy or support (or perhaps, your supportive attitude). Thus if you translate *misericordia* as "pity," it sounds somehow bad or condescending, at least in part. But if you translate it as "sympathy," it sounds much better. These connotations have important implications for how argumentum ad misericordiam is perceived as a fallacy. For if you translate the name of the argument as "appeal to pity," it immediately creates a strong presumption that this type of argumentation is inherently fallacious or wrong.

This negative aspect is very well brought out in the entry for 'pity' in the *Dictionary of Philosophy* (Runes 1964, 236).

> Pity: A more or less condescending feeling for other living beings in their suffering or lowly condition, condoned by those who hold to the inevita-

bility of class differences, but condemned by those who believe in melioration or the establishment of more equitable relations and therefore substitute sympathy (q.v.). Synonymous with "having mercy" or "to spare" in the Old Testament (the Lord is "of many bowels"), Christians also are exhorted to be pitiful (e.g., 1.Pet.3.8). Spinoza yet equates it with commiseration, but since this involves pain in addition to some good if alleviating action follows, it is to be overcome in a life dictated by reason. Except for moral theories which do not recognize feeling for other creatures as a fundamental urge pushing into action, such as utilitarianism in some of its aspects and Hinduism which adheres to the doctrine of *karma* (q.v.), however far apart the two are, pity may be regarded a prime ethical impulse, but due to its coldness and the possibility of calculation entering, is no longer countenanced as an essentially ethical principle in modern moral thinking.—*K.F.L.*

Here the difference between sympathy and pity is brought out very clearly. 'Pity' has negative connotations of a condescending attitude, even suggesting elitism, whereas 'sympathy' is "substituted" by those who believe in equality. The "appeal to sympathy" sounds very positive and nonfallacious, whereas "appeal to pity" has all the negative implications brought out above and is easy to condemn as a "fallacy." The other available words, 'compassion,' 'mercy,' 'charity,' and 'tender-heartedness,' sound more old-fashioned, but are also probably less objectionable than 'pity.'

Some argumenta ad misericordiam are reasonable, that is, nonfallacious (Walton 1992b, 112–19). A typical case of this type is a charitable appeal for action to aid victims of suffering. Typically this type of appeal to sympathy uses argumentation from example, by presenting a specific example of some person who is suffering. Then the appeal uses argumentation from consequences, arguing that the bad situation could be made better by some action that the respondent could take.

A full-page ad in the *Sciences* (March/April 1991, 29) has a large picture of a small child (Mehelina from Bolivia, age six) who is crying or at any rate looks distressed. The main part of the ad reads:

> *Case 105*
>
> If you met a child like Mehelina in Bolivia, so malnourished she's too weak to laugh or play . . .
>
> or Roberto in Colombia, at work in the fields instead of school, too tired to hope or dream . . .
>
> your heart would break. Your sense of what's right would be outraged. You'd want to help. But how?
>
> As a Childreach Sponsor. It's a way to help that's *simple, personal— and effective!*
>
> A way to care. To connect.

To create real change.

As a Childreach Sponsor, you *can* reach out to a child who really needs you. Not only with donations, but with your caring and encouragement. Because you believe in each child's value in the Family of Man.

It only takes $22 a month to provide hope—and help—like better nutrition, health care, and education for the child. New income raising skills for the family. And clean water for the community.

The ad closes with instructions on how to become a sponsor of a child, concluding with the exhortation (in large letters): "Yes, I want to reach out and make a difference." This type of ad is very common, and the reader will be familiar with the type of argumentation it uses.

As we all know, it is not easy to ensure that your charitable donation actually has the effect of helping someone, and nowadays charitable agencies are listing their "administration" costs. We know as well that such appeals can be a "scam" based on entirely false claims in some cases. Even so, in principle, the use of argument to sympathy in presenting the details of an actual example in a charitable appeal of this sort can be a reasonable (nonfallacious) type of argument.

The type of argument involved is a use of practical reasoning involving argumentation from consequence to press the respondent for action to alleviate a bad situation. The argumentation scheme below could be called the *argument from supplication*, a kind of plea for help.

Individual *a* is suffering.
Bringing about *A*, some form of help, will alleviate *a*'s suffering.
Therefore you, the respondent, *b*, should bring about *A*, if you can.

The first premise of this sequence of practical reasoning presents a particular case or example that appeals to the empathy of the respondent. The second premise is a promise, and the conclusion is a request that recommends a particular course of action. Further parts of the argumentation typically give specific directives on how this means can be implemented (by sending money or something of that sort) by the respondent directly.

Another typical type of case where the argumentum ad misericordiam can be used reasonably to shift a burden of proof in a dialogue is the kind of argument like that in case 80a, where an excuse is put forward to argue that an exception should be made to a rule. The notorious problems with using the type of argumentation, however, are graphically illustrated by cases 9, 10, and 11. Sometimes such arguments are based on very poor excuses, but the proponent tries to embellish the case by emotional appeals to pity that may not even be relevant.

In case 11, the main critical failure was that the premise was simply false. As it turned out, the claim that babies were pulled from incubators was a fabrication, with no evidence to substantiate it at all. The problem here is whether to classify the argument as a fallacious appeal to pity or simply as an appeal to pity based on a false premise. The evidence for calling it a fallacy is that evidently the whole scheme was a public relations tactic carefully engineered with the help of a sophisticated public relations company that had close links to the Kuwaiti royal family. A systematic tactic of deception was involved, and this inclines us to call the argument a fallacy rather than just an argument with a false premise.

In cases 9 and 10, the problem is one of relevance. The proponent is using the appeal to pity to change or twist the issue of the dialogue.

9. The *ad Populum* Fallacy

Many examples of the ad populum fallacy cited in the textbooks are not fallacious arguments but only weak arguments that give just a small weight of presumption that needs to be evaluated against a larger weight of evidence. For example, following political polls or the party position on an issue is often cited as the ad populum fallacy, as indicated by this case (Damer, 1980, 90).

> *Case 106*
>
> The fact that the platform of the Republican party supports an anti-abortion amendment to the U.S. Constitution does not constitute a good reason for Senator Davis, a six-term Republican senator, to support that amendment. Loyalty to one's political affiliates should not play any significant role in the formulation of one's position on such an issue. Hence an appeal based upon that consideration would probably be an irrelevant one. A proper appeal would focus on possible reasons such an amendment might be needed.

What should be said here is that in a democratic system, an ad populum argument based on perceived majority sentiment, or on a party position, should actually be taken generally as a presumptively reasonable kind of argumentation that carries some weight but is defeasible if overridden by stronger evidence in a case.

Where the fallacy comes in is in the kind of case where the inherently weak and defeasible ad populum argument is given too much weight, and other relevant evidence that would counterbalance it is ignored or even systematically excluded from consideration. This counterbalancing aspect of the ad populum argument in democratic politics has been expressed nicely by Samuelson (1992, 51), who describes the politician's job. "Their hard task is to maintain a crude balance between popular pressures and a larger concept of national interest. Our predicament today is that the balance has been all but lost. A healthy respect for public opinion has become a slavish devotion, motivated by the desire to be re-elected." It is this failure of balance that puts too heavy a weight of presumption on perceived popular opinion, based on polls, that makes the ad populum argument such a common fallacy in political argumentation.

In principle, we must remember that the ad populum is a reasonable presumptive type of argumentation that can rightly be used to shift a burden of proof in a dialogue, as indicated by the argumentation scheme for the argument from popular opinion given in chapter 5, section 8. So we can't declare all instances of it fallacious. Instead, each case must be examined on its merits. Of key importance is the second critical question for the argumentation scheme for the argument from popular opinion, which asks whether counterbalancing relevant evidence is available in the given case.

The argumentation scheme for the argument from popular opinion is in fact only a general template or stencil that needs to be specified or filled in by making explicit what type of "reference group" is meant. Depending on which reference group is meant, several subtypes of ad populum argument can be specified.

1. *Argument from Popularity.* This is sometimes called the "bandwagon argument." Freeman (1988, 71) calls it *bandwagon appeal*, which occurs where it is concluded that a belief *must be* true or an action *must be* right because most or all people accept it or approve of it. Freeman (71) describes this argument as a fallacy because "popularity is a reason, but a weak reason" for accepting a conclusion. Similar accounts are given by Johnson and Blair (1977, 159), who call it "popularity," and Hurley (1991, 114), who calls it the "bandwagon argument." See also Walton (1989a, 89) on the basic form of the argument from popularity.

2. *Mob Appeal.* What is essential here is the appeal to emotions or "enthusiasms" of the crowd. Engel (1976, 113) describes "mob appeal" as "an argument in which an appeal is made to emotions, especially to powerful feelings that can sway people in large crowds." Here, presentation in a "theatrical manner" is important. Hurley (1991, 113) calls this type of ad populum argument "the direct approach," which "excites the emotions and enthusiasms of the crowd," citing propagandists and demagogues like Hitler as offenders. This makes the argument sound pretty bad. Copi and Cohen

(1990, 103) also cite as a "classic example" the "speeches of Adolf Hitler, which brought his German listeners to a state of patriotic frenzy."

3. *Appeal to Fashion.* Copi and Cohen (1990, 103) hit heavily here on the advertising agencies, "ballyhoo artists" who "sell us day-dreams and delusions of grandeur" by associating their products with people or things that we approve of or which "excite us favorably." This type of argumentation could also be called appeal to snobbery, vanity, trendiness, or "what's in" (or perhaps there are subclassifications to be made here). Hurley (1991, 114) calls this an *appeal to vanity*, which associates a certain product with a "celebrity who is admired and pursued."

4. *Position to Know of Group.* In some cases, arguments from popular opinion are given weight because the reference group cited is in a special position to know or even includes experts of some sort. In case 87, for example, Karen and Doug were strangers in a foreign country, and lacking personal knowledge of local customs, they presumed that because everybody else was riding side by side, it was okay to do it. Here we can see that the ad populum argument is mixed in with position-to-know argumentation and with the ad verecundiam fallacy as well, to some extent. Freeman (1988, 71) explicitly notes this connection when he connects "appeal to the glamorous person" with appeal to authority in cases where movie stars or popular musicians are used to promote products or causes.

5. *Plain Folks Argument.* Typically, in this type of argument, the political speaker portrays the opposition as "elitists" and tries to portray himself as "an ordinary guy." A classic case was the 1988 Canadian federal election debate (Walton 1992b, 83–85) where in the midst of the debate on free trade, Mr. Mulroney described how his father "went himself, as a laborer, with hundreds of other Canadians" to build a "little town" in Quebec. Not to be outdone, Mr. Turner replied that his mother was a miner's daughter in British Columbia. Mr. Mulroney used his argument to conclude "I love Canada" and Mr. Turner countered this by saying Mr. Mulroney had caused us (Canadians) to be in danger of becoming a colony of the United States.

6. *Rhetoric of Belonging.* In this type of ad populum argument, the speaker reinforces solidarity with his audience, suggesting that anyone who disagrees is excluded from the group. The presumption is that anyone who isn't for the cause can be excluded from the group and is therefore not worth listening to. This type of tactic is a kind of closing off of critical discussion by shifting to a quarrel or negotiation dialogue. A classic case is the speech of Walter Reuther (Walton 1992b, 100–101) where, as Bailey (1983, 134) pointed out, nonbe-

lievers are excluded as people who "want to make a fast buck." These could be called *insider arguments* because they make inclusion or acceptance in the group as a requirement of acceptance of any argument in the dialogue exchanges.

7. *Common Consent.* With many reasonable ad populum arguments, there is some existing framework of dialogue in the given case that makes it a basis of entering into the dialogue that consent of the participant, or of some designated group, will be a criterion of acceptance of a conclusion. In a democratic election, for example, it may be agreed that the majority will decide by a vote or referendum. In a legal trial, it may be part of the accepted framework of dialogue that a jury will decide the outcome of a disputed case.

8. *Peer Pressure Argument.* In some cases, the reference group cited in the argument from popular opinion is composed of persons who are in the same situation or group. The argument is to the effect that if they are allowed to do things in a certain way, and if I am not, then I am being excluded or treated unfairly. This argumentation is commonly used by children in pleading with parents and is a very familiar kind of tactic.

9. *Moral Argument.* Another subtype of ad populum argument is the appeal to popularly accepted practices or ethical standards to argue for a vindication, permissibility, or even rationalization of one's own allegedly culpable actions. A classic case was the response of looters in the Los Angeles riot to interviewers who responded, when their actions were questioned, "Everybody is doing it." Some young looters interviewed admitted that what they were doing was wrong but used this argument as an excuse.

10. *Consensus Gentium* (Consensus of the Nations). This variant is based on the premise that all the civilized nations do something in a certain way, or have adopted a certain belief. The conclusion indicated is that this belief or course of action must be right. The traditional case is the common consent argument for the existence of God (Edwards 1967, 147–55). A simpler example is case 8, where it was argued that every civilized country in the world has done away with capital punishment. Typically, in this type of argument, common consent or practice is not the only basis of the argument but some additional reason; for example, being "civilized" is added to back up the evidential value of the common consent.

To sum up, we can see that the ad populum argument, as actually used, is not unproblematic to define precisely, because it admits of these numerous variants or subtypes. What they have in common, as ad populum arguments, is the argumentation scheme for the argument from popular opinion given in chapter 5, section 8.

This type of argumentation is in principle reasonable, or nonfallacious, as a species of presumptive argument used to shift a burden of proof in a balance of considerations dialogue where hard evidence does not resolve the issue. It becomes fallacious when we lose sight of its inherently presumptive and defeasible nature and take it for a more decisive argument than it really is. It can also be a fallacy on grounds of failure of relevance, for example, in a scientific inquiry, where the issue should be decided only by hard (nonpresumptive) evidence.

10. Toward a Theory of Fallacy

We have made a good foundational beginning at analyzing six of the major informal fallacies, using the tools of the argumentation schemes and the types of dialogue. The problem now poses itself of whether we can generalize from this work, as tentative and incomplete as it remains at this stage, so as to construct some hypothesis about the concept of fallacy. For this is the other tool we need, if we are to make progress in analyzing and evaluating the fallacies.

Already some hypotheses have been ruled out. We can't simply say that a fallacy is an argumentation scheme that the argument in question fails formally to conform to or to be an instance of. This approach is least plausible for fallacies like many questions and begging the question, where longer sequences of argumentation, or argumentation themes, are involved. But it doesn't even work, at least directly, with the other four fallacies studied, even though these fallacies do closely relate to argumentation schemes, and the scheme is an essential ingredient in their analysis and evaluation.

Also, we can't simply say that a fallacy is a violation of a rule of dialogue, like a critical discussion. While this is a necessary condition of something's being a fallacy, it is not by itself sufficient, nor does it give us an analysis of the concept of a fallacy that defines essentially what a fallacy is, as a basic concept of informal logic. There simply is no one-to-one correspondence between the individual fallacies and the types of violations of rules for a critical discussion. Moreover, this definition overlooks the important distinction between fallacies and other types of rule violations of a less serious sort, such as blunders and inadequately supported arguments.

A better approach needs to begin with the recognition that the types of arguments associated with the fallacies are, in some instances, reasonable (nonfallacious arguments). But even before this hypothesis can be absolutely confirmed, though the evidence for it has already been shown to be abundant, the problem of defining the individual fallacies as distinctive types of argumentation has to be solved. This recognition too militates against defining a fallacy as a violation of any single rule of a dialogue. For each fallacy can, and typically does, involve more than one way of failing to be a good argument. And moreover, there are subtypes of each type of argument coming under the heading of each fallacy.

With the ad hominem, each of its three main subtypes, the direct, the circumstantial, and the bias type, can be a presumptively correct argument based on its characteristic, defining argumentation scheme. And each subtype can fail in one or more of three ways—a premise may be inadequately supported, there can be a shift from presumptive raising of critical questions to an absolutizing of the conclusion as false, or there can be a failure of relevance. Thus there is no single type of failure that can be identified (on a one-to-one basis) with the ad hominem fallacy. We must, it seems, take a broader approach to defining the concept of fallacy, in relation to ad hominem. It is a complex of different types of argumentation, all of which share the defining characteristic of being types of personal attack used to question or discredit an opponent's argument but each of which can go wrong or be fallaciously used in different ways. To come to know the fallacy, one must catalog the different ways this complex of argument types can be used wrongly. All of these ways tend to be highly contextualized, so that evaluating in a particular case whether the ad hominem fallacy has really been committed is best done by constructing a profile of dialogue (Walton 1985a). Moreover, evaluating such a profile involves a careful analysis of what type of dialogue the participants were supposed to be engaging in and being alert to dialectical shifts. Much the same lesson turned out to be true of the ad baculum fallacy, where the existence of an underlying dialectical shift is the key to analysis and evaluation of many of the most powerful and deceptive cases of this fallacy.

Both ad misericordiam and ad populum share the characteristic with ad hominem of being a fallacy in a given case for two main reasons. One is failure of relevance, and the other is overestimation of the strength of the argument as support for its conclusion. With ad hominem this overestimation of strength can take two forms—the premise can be insufficiently supported, or there can be an absolutizing of the conclusion.

In light of chapter 6, we can appreciate generally how failure of relevance can be a dialectical fault associated with a fallacy. With these four *ad* fallacies, because of their emotional impact (Walton 1992b), failure of relevance is easily disguised or overlooked in the heat of a debate even more easily where there is a shift to the quarrel.

But with overestimation of strength as a defining characteristic of a fallacy, more care needs to be taken. For a weak argument—I mean an argument inadequately supported by evidence—is not necessarily a fallacious argument. What is the difference between a weak argument (in this sense) and a fallacy? And what is it about some weak arguments—for example, abusive ad hominem attacks based on "smear tactics"—that makes them such powerful sophistical tactics that, in some cases, they are called fallacies, even if they are relevant? This difficult but central question calls for a more careful analysis of the concept of a fallacy. Before reading the details of this analysis in the next chapter, it will be evident to the reader that a new approach to the concept of a fallacy is being advocated. No longer can we take it for granted that the traditional fallacy labels always mark cases of arguments that are (in our sense) genuinely fallacious. Instead, such a question is a judgment or conclusion that requires evidence to back it up.

According to this new approach, any claim that a fallacy has been committed must be evaluated in relation to the text of discourse available in a given case. The first task is to locate the argument, that is, generally a set of premises and a conclusion, that supposedly contains the fallacy. Such an argument will always occur in a context of dialogue, according to the new theory. Much of the work of analysis and evaluation of the allegedly fallacious argument will involve placing that argument in a context of dialogue.

When we deal with fallacies, there are generally two parts or aspects of a given argument to be concerned with. First, there is the argumentation scheme, or form of the argument. For example, if it is a formal fallacy based on a deductive type of reasoning, the form could be that of modus ponens. Or if it is an informal fallacy based on argumentation from consequences, the argumentation scheme could be of the type for argumentation from consequences.

Evaluating the argument at this first level, it can be criticized on two grounds. First, it may be an invalid or structurally incorrect argument or may otherwise fail to conform to the requirements appropriate for that type of argumentation scheme. Second, one or more of the premises can be criticized on the grounds that it has been inadequately supported. In the case of presumptive argumentation schemes, this means that appropriate critical questions have not been adequately answered.

The second aspect of an argument to be considered is that of relevance. Even if the argument is a good one at this first level, it could still fail to be relevant. But what do we mean here by 'relevant'? It has been shown in chapter 6 that relevance, in this sense, is dialectical relevance, meaning that an argument is relevant if it fulfills its proper function in the given context of dialogue.

At the first level of criticism, one is mainly concerned with the premises and conclusion of the argument—what might be called its inferential structure and content. This could be called a *local* level of analysis, because the concern is with the premises and conclusion of a single argument rather than with the broader use of the inference in a context of dialogue.

Where a major informal fallacy has occurred in a given case, in some instances the error can be revealed, analyzed, and evaluated at the local level as an error of reasoning. We might look at the argumentation scheme, for example, and point out that one of the premises has not been adequately supported. The problem here may simply be an error or oversight, but if it is a serious enough one (in a sense to be analyzed in chapter 8), we may rightly say that a fallacy was committed.

In other cases, however, things may not be this simple, because the fallacy can only be documented and proved by bringing forward textual evidence to show that the arguer's use of the argumentation theme over a protracted sequence of dialogue reveals an uncooperative, tricky, or deceptive use of argumentation. Such a use of sophistical tactics can be evaluated as fallacious because, or to the extent that, it blocks the legitimate goals of the dialogue that the participants in argumentation are supposed to be engaged in. This blockage can be revealed in a profile of dialogue.

In this type of case, we must take a broader, pragmatic view of the concept of fallacy that takes the dialectical context of an argument more deeply into account. Instead of looking at the argument mainly at the local or micro level, we must make a more serious effort to evaluate systematically the larger context of dialogue in which the argument was used.

8

A Theory of Fallacy

A major problem that frequently appears in attempts to analyze and evaluate fallacies is the concept of fallacy itself. What does it mean to say that an argument is fallacious? In deductive logic, we have the decisive advantage of being able to distinguish between invalid arguments and arguments with false (or inadequately supported) premises. And we have forms of argument, like modus ponens, to back up such evaluations on some kind of stable basis of evidence.

The problem with informal fallacies was that we never had comparable kinds of exactly defined structures, or abstracted counterparts to the given forms of argument, on which to base our judgments that such-and-such an argument is fallacious or not. Now we have the argumentation schemes of chapter 5 and the types of dialogue in chapter 4. But how do these structures fit together in a framework in which we can usefully define the kind of failure of argument traditionally known as a fallacy?

The basic problem is that tradition is ambivalent and confusing here, even (Hamblin 1970) somewhat disorienting. The ancient idea of fallacy in Aristotle's writings on the subject was a "sophistical refutation"—a deceptive trick of argumentation used to get the best of a speech partner in a dialogue. This idea was never really taken seriously by generations of subsequent logicians who, building on Aristotle's syllogistic (deductive) type of logic, lost sight of the concept of an argument as an interpersonal exchange where two parties reason with each other. For them, the ancient idea of fallacy was simply not comprehensible as part of the science of logic mathematized by Boole and Frege.[1]

1. What a Fallacy is Not

In English the term 'fallacy' has come to have a very broad meaning. It includes not only misapplied argument techniques but also errors, misunderstandings, or false beliefs of widespread appeal. Under the term 'fallacy,' English can include a widely held or accepted belief that is nevertheless false even if this belief is not the conclusion of an inference or argument. In a reference book, *Popular Fallacies* (Ackermann 1970), many beliefs that are popular, or were at some time or other, are listed as fallacies along with a short paragraph citing authoritative scientific or historical sources that show these beliefs to be false. For example, the following items are listed as fallacies.

That a Man has One Rib Fewer than a Woman[2]
That Lightning Describes a Zigzag Path with Acute Angles[3]
That Eunuchs were Introduced into Europe by the Turks[4]

In this usage, the word 'fallacy' does not necessarily refer to misuses of argumentation techniques or failures of logical inferences. Instead, for the most part, the "fallacies" cited seem to be popular beliefs that do not have a factual or scientific basis. They appear, for the most part, to be false beliefs based on folk wisdom, popular culture, or superstitions. They are not instances of types of baptizable failures of logical reasoning that should come under the heading of informal fallacies, or any other (e.g., formal) kind of fallacies, as the subject of logic is, or should be taught as a systematic method of argument evaluation.

A basic requirement of a fallacy, in the sense of the word appropriate for logic, is that there must be a fallacious argument or at least a structural failure in something that was supposed to be an argument as used in a dialogue where argumentation is supposed to be taking place. This is called the *argument requirement*, and the exact form it should take becomes a subject of controversy in chapter 9. In chapter 8, we will operate on the presumption that the argument requirement, in some form, is part of the concept of a fallacy.

We should also distinguish between fallacies and other kinds of less serious errors in arguments, like blunders and *lacunae*, or gaps. A blunder can involve breaking a rule of reasoned dialogue, but a fallacy is a more serious kind of infraction that involves a systematic technique of deceptive argumentation. For example, in some cases, a participant in a discussion may inadvertently argue in a circle, resulting in a weak and unpersuasive argument that would fail to convince anyone. Such a blunder may not be a case of committing the fallacy of begging the question, according to chapter 7, section 3. To commit the fallacy of begging the question is a serious matter that involves an aggressive attempt by one participant in a context of dialogue to use a circular sequence of argumentation to try to convince another participant erroneously and misleadingly that he (the first participant) has properly met the burden of proof appropriate for this context (Walton 1991a). Thus to commit this fallacy involves more than just inadvertently arguing in a circle. Indeed, in some contexts of argument, circular argumentation is not fallacious but is characteristic of feedback reasoning, a legitimate form of practical reasoning when deciding on a course of action in relation to a knowledge base of particular circumstances, as we found in chapter 7, section 3.

We have seen already that there is an important class of distinctions between the concept of a fallacy and other weaker classifications of criticisms of errors in arguments. If an argument contains a flaw, gap, hole, or weak point, it should be subject to critical questioning on this point. That is one type of criticism. But if an argument commits a logical fallacy, then it is open to a much stronger form of criticism, requiring that the argument contain a flaw that not only is a structural failure of the reasoning it contains but also is dangerous to the talk exchange of which it is a part.

It is quite a serious criticism to allege to someone in discussion that her argument "commits a fallacy." In English, this form of rejoinder is a serious kind of censure or reproof that borders on the impolite. It suggests that the person so reproved is generally basing her thinking on some underlying systematic confusion or error based on her misunderstanding or ignorance of the principles of reasoning. This is strong stuff. It is much more usual, and polite, to criticize someone's arguments, or moves in argument, by raising questions about the argument and its assumptions, which the person criticized can then reply to. To criticize someone's argument by saying that she "committed a fallacy" leaves the alleged offender no way out for polite rejoinders. She must either contritely admit her error or attack the charge with vigor.

For this reason the term 'fallacy,' as the leading, general term for important kinds of techniques and strategies of argumentative discussion that can lead to problems, confusion, or deceit, is not always the most felicitous term in English. The term 'fallacy' is sometimes appropriate to describe this kind of error, but in many cases, this term is too strong. It is often more appropriate, for example, to speak of arguments or argument strategies that are flawed, weak, or poorly presented or that suffer from specific gaps or shortcomings that can be clarified or corrected in subsequent dialogue.

The confusion and problems inherent in the current textbook usage of the term 'fallacy' (Hamblin 1970) suggest that the classification of errors, weaknesses, and tricks of argumentation in dialogue ought to be systematically rethought and that a more refined definition of the concept of a fallacy ought to be considered.

At the monological level characteristic of traditional deductive logic, an argument can be evaluated as *weak* if the premises are not adequate to prove or establish the conclusion according to the required standard. By contrast, an argument can also fail to be deductively valid. If deductive validity is the required standard, then the argument that fails to meet the standard can be declared weak or inadequate. Various standards are possible here in different contexts of argument. For example, inductive strength (which could possibly be specified precisely in different ways) might be another standard. This is the familiar framework of argument evaluation that has dominated logic for two thousand years.

But now it is appropriate to bring in a new framework of evaluation of applied logic. At the dialectical level, an argument used as a move (or sequence of moves) in dialogue can rightly be evaluated as a fallacy if it twists some scheme or theme of argument rightly used in some context of dialogue to the advantage of the participant who has made the move or sequence of moves in (possibly another) context of dialogue. By contrast, a *blunder* is a move that breaks a maxim of dialogue but is not the inappropriate use of a technique of argumentation to promote the goals of the mover deceptively. This account of 'fallacy' presupposes that in a context of dialogue there is some set of maxims of the dialogue that prescribe how and where appropriate moves or sequences of moves should be made—a kind of code of guidelines for cooperative and reasonable discussants. It is the thesis of van Eemeren and Grootendorst (1984, 177) that fallacies can be regarded as violations of rules of dialogue that formulate a code of conduct for rational discussants. Because a fallacy ought to be regarded as quite a serious violation, however, in light of our remarks on the English usage of this term above, room should be made for less serious violations or flaws as well. Blunders can be serious as well, but they are not as strategically interesting to study as fallacies, because the latter are systematic strategies, moves, or patterns of moves that are instruments important in constructing systematic patterns of attack and defense in argumentative dialogue. Fallacies are species of violations of a rule of a dialogue, but that condition is not sufficient to define them as fallacies.

Knowing whether a sequence of moves in argumentation is a fallacy, as opposed to a blunder, involves knowing the context of dialogue in order to know which goals, techniques for realizing these goals, and rules are appropriate. But it also involves defining that type of fallacy as a characteristic technique or sophistical tactic. A circular argument that is a mere blunder in one context of dialogue could be a vicious circle, a fallacy, in another case where the rules, methods of proof, and requirements of the discussion are different.

Because of the incompleteness and problems of interpretation with particular texts of discourse, there are cases where it may require assessing a lot of textual evidence to judge whether a given argument is better classified as a fallacy or as a weak but nonfallacious argument. In a typical case of this type, an argument may be an instance of a general technique associated with a fallacy—for example, it may be an ad hominem argument—but the error committed does not seem serious enough to justify calling it a fallacy as used in this particular instance. It may seem more like a blunder than a fallacy in this instance.

A good example is the use of the circumstantial ad hominem argument in an instance where a presumptive case for a pragmatic inconsistency is made but the specific links between the arguer's commitment and the action alleged to be contrary to it are not spelled out in sufficient detail. Such a case may be a weak ad hominem argument that is open to critical questioning, but it may not be such a bad argument that it merits being called fallacious. For the argument may be partly reasonable, to the extent that it does make enough of a case to shift some burden of presumption against the party accused of "not practicing what he preaches." The modern textbooks, under the philosophy of routinely classifying any argument that "seems valid but is not" as a fallacy, tend in the direction of calling such cases "fallacies" far too often. This tendency should be restrained by more carefully examining each case sufficiently on its merits.

We should resist acting precipitously in this type of case, if the error in the case is not a serious enough one to justify calling the fault a fallacy. Even though the argument in the case in point is an instance of a type of argumentation generally classified as a fallacy by tradition, we should resist categorizing every instance of it as a case where a fallacy has been committed.

Our new approach entails rethinking the concept of fallacy. Just because a particular argument embodies a technique of argumentation that can be used fallaciously in some context or other, it does not follow, on the new approach, that the particular argument is fallacious. On the new approach, there is a logical gap between these two findings in a given case.

2. Six Basic Characteristics of Fallacy

The concept of fallacy is inherently negative. A fallacy is a bad thing—fallacies are to logic as pathology is to medicine. A fallacy is a kind of failure or negation of correct argumentation. In the new theory, which defines a fallacy as an argumentation technique used wrongly, this negative aspect is reflected in three basic characteristics of fallacy. A fallacy is paradigmatically three things.

1. A failure, lapse, or error, subject to criticism, correction, or rebuttal.
2. A failure that occurs in what is supposed to be an argument (argument requirement).
3. A failure associated with deception or illusion.

According to the first characteristic, when a fallacy is "committed," its commission is rightly subject to some form of censure or condemnation. A fallacy is not just any error, false belief, rudeness, trick, or impropriety in dialogue, however. Fallacies are associated with the wrong use of argumentation schemes.

It follows that fallacies occur in arguments or at least in a context of dialogue where there is supposed to be an argument. It does not follow that every instance of a fallacy is an explicit argument of any of the more familiar types in traditional deductive or inductive frameworks. As we saw in chapter 7, many fallacies involve the use of emotional distractions, like threats, for example, that need not be (in themselves) arguments that are reasonable or correct forms of argument in a critical discussion. But in such cases there is a type of argumentation that has been used wrongly—for example, a tactic of irrelevance used for distraction or evasion in one type of dialogue—but that can be recognized as an argumentation scheme appropriate for use in another type of dialogue.

The third characteristic can be identified with the idea that the argument "seems correct." But this characteristic should not be interpreted psychologically to mean that the offending argument must always seem correct to the person to whom it was directed, either in every case or in any particular case. What is important here is the idea that certain common techniques of argumentation are powerfully effective in a context. Used in the proper context of dialogue in the proper way, such techniques are powerful because they are reasonable ways of carrying out argumentation. There can be shifts of dialogue underlying the deceptive misuses of such techniques, however. A normally reasonable kind of argument is perverted, when such a shift takes place, if the shift is concealed, not bilaterally agreed to, or otherwise illicit. But even so it may seem correct because we are normally accustomed (in another familiar context) to accepting this kind of argumentation. The "seeming correct" aspect is therefore a matter not of psychology but more of reasonable expectation and presumption based on argumentation schemes as standard routines in contexts of argumentative dialogue. Fallacies work because they take advantage of standard expectations in dialogue, where there is a normal presumption on the part of each participant that the other participants will collaboratively follow maxims of politeness. Violations of these expectations can be subtle and deceptive when a shift is concealed.

Although the most worrisome cases occur where one party in a dialogue is deliberately using a tactic of argumentation as a trick to get the best of another party unfairly and deceptively, fallacies can also occur where the seeming correctness does not spring from an intent to deceive. Using a technique wrongly, even where it is highly inappropriate and strongly contrary to the aims of reasonable dialogue, does not imply an intent to deceive, in every case, by one party in the dialogue.

Three further characteristics round out the new pragmatic concept of fallacy.

In the definition of a fallacy as a technique of argumentation that is used wrongly, the phrase 'used wrongly' means three things, referring to:

4. A violation of one or more of the maxims of reasonable dialogue or a departure from acceptable procedures in that type of dialogue.

5. An instance of an underlying, systematic kind of wrongly applied technique of reasonable argumentation (argumentation theme).

6. A serious violation, as opposed to an incidental blunder, error, or weakness of execution.

Taken together, these three essential characteristics of fallacy imply that advancing a charge of fallacy is a serious kind of accusation in argument that carries with it ramifications for all participants involved in the dialogue. On the part of the accuser, this charge carries with it a burden of proof to back up the accusation with sufficient evidence of the right kind to make the charge stick. On the part of the accused, this type of charge is a serious indictment that calls for a strong and vigorous response in rebuttal. A failure to reply with adequate strength carries with it a presumption that the accused has committed a rule violation and concedes the point of contention at issue in the charge.

A fallacy is said to be "committed," implying an error, transgression, or mistake that needs to be corrected or rectified. Once a fallacy in an argument is identified, or pointed out, the argument needs to be corrected or withdrawn. That is, the required response to any particular instance of a fallacy is to point out the speech act in question as an erroneous move in argumentation and to demand its retraction or correction. A fallacy, then, is a move in argument that requires a normative response identifying it as an illicit move not to be allowed as a correct argument move.

The charge "You have committed a fallacy here" invites a response by the would-be offender who is challenged. The charge alleges that an error has been committed, and the arguer charged has a burden of making his error right. Failure to fulfill this burden carries with it an implicature that the argument containing the error is of no further worth as a contribution to the discussion.

The concept of an erroneous move in argument, in this context, implies that there are underlying procedural maxims of dialogue governing a discussion and that a fallacy is, or involves, a deviation from one or more of these maxims. A fallacy is an instance of a breach of the standard of what is an appropriate purposefully contributory way of arguing in a reasoned dialogue. Therefore, a fallacy is basically a wrong use of a right procedure of argumentation in a particular type of dialogue.

In principle, it is a necessary (but not sufficient) condition of something's being a fallacy that it be a violation of a rule of reasonable dialogue. But not all reasonable dialogue is rule based. Some cases of dialogue are better treated as *frame based*, meaning that although (all) the rules are not stated explicitly, certain types of procedures are expected as usual routines of argumentation, techniques that are generally accepted in practice. In frame-based dialogues, it is presumed that argumentation will conform to maxims indicating these generally acceptable procedures. The presumption generally is that any given sequence of argumentation is an instance of one of these familiar procedures so often used, unless there is some reason for suspecting that it is or may be departing from the usual way the procedure is run.

A frame is like a rule except that it is not stated in such an explicit and rigid way. Frames admit of exceptions, and the way a particular case is judged to be an exception or not is based more on familiar expectations about what sort of practice one can reasonably expect in a particular type of case. The notion of a frame is inherently pragmatic because it is based on accepted practices of what arguers experienced in dealing with a particular type of exchange can reasonably expect of each other. These expectations are expressed in maxims (Grice 1975) that create defeasible presumptions of correct conduct, corresponding to expectations of politeness in conversation.

How one can tell that a systematic kind of wrongly applied argumentation technique is a fallacy in a given case is likely to involve not only citation of a rule violation in many cases. One has to look at the given sequence of argumentation and specify how it has supposedly gone wrong or departed from acceptable conversational standards of politeness.

For example, is attacking someone's character in an argument fallacious or not? It all depends on how it's done in most cases. In some contexts of dialogue—for example, in a scientific inquiry—attacking a participant's character would clearly be against the conversational maxims altogether and would be ruled out of the dialogue. But in other cases—for example, in a political debate—it might not be so easy to exclude argumentation that attacks someone's character purely because it is a character attack per se. Instead, it is a question of what the maxims of conversation can tolerate as appropriate in this particular case. How far has the attack gone beyond the normally accepted practices of pressing an argument forward? Personal attack may not be wholly or incontrovertibly against the rules. The problem is better posed as a question of how the use of personal attack as a technique for refuting an opponent's argument has been rightly or wrongly applied in the dialogue in this case. It could be a legitimate use of the technique. It could be a badly executed use of the technique—open to critical questions and rebuttals on various counts. Or worst of all, it could deviate so strongly from proper use of the technique that it impedes the dialogue seriously enough to be called a fallacy.

3. A Fallacy is an Illusion or Deception

In addition to being an error, a fallacy has traditionally been regarded as a kind of error that is based on an illusion of correct argument. Thus a fallacy is a kind of counterfeit—something that appears, on the surface, to be genuine but underneath is a fake. A fallacy is not what it appears to be, and this misleading appearance of being genuine makes it dangerous and confusing in arguments. Hamblin (1970, 109) noted that the word *fantasia* was sometimes used, in the Middle Ages, as a synonym for *fallacia*.

Aristotle conveyed the idea of fallacy as illusion very vividly in the *De sophisticis elenchis*. At the very beginning of this treatise on the fallacies, he contrasted reasonings that are "really reasonings" with those that "seem to be, but are not really, reasonings" (1955, 164 a 23). To make this contrast vivid, Aristotle drew comparisons with other illusions based on false appearances.

> Some people possess good physical condition, while others have merely the appearance of it, by blowing themselves out and dressing themselves up like the tribal choruses; again, some people are beautiful because of their beauty, while others have the appearance of beauty because they trick themselves out. So too with inanimate things; for some of these are really silver and some gold, while others are not but only appear to our senses to be so; for example, objects made of litharge or tin appear to be silver, and yellow-coloured objects appear to be gold.[5]

According to Aristotle, reasoning and refutation are like this as well—sometimes real and sometimes illusory. When they only appear to be real, it is because of "man's inexperience" (1955, 164 b 27). Hence men can become victims of false or merely apparent reasoning, and the sophistic arts can take advantage of this vulnerability.

It was perhaps this Aristotelian emphasis on fallacies as reasonings that are merely apparent rather than genuine that led to the glib definition of *fallacy* found in most modern logic textbooks—an argument that seems valid but is not. But defining a fallacy as an argument that is not valid but seems to be valid is not good enough. To evaluate a particular argument as a fallacy, you must look to the context in which it was used in a particular case. This task is a matter not purely of identifying a propositional logical form that the argument fails to have but of looking to how the argumentation theme or scheme was put forward as something that was supposed to meet a standard of correct use in a dialogue. Then you must diagnose how it departed from that standard of correct use by constructing two parallel profiles of dialogue.

But an even more significant error is to be found in the presumption that the sophistical tactics types of fallacies are incorrect or faulty instances of reasoning just because they are instances of semantically invalid structures of logic. For the most significant and dangerous sophistical tactics fallacies in natural language argumentation—those outlined in chapter 3 and the ones so far analyzed in chapter 7 in this book—are dialectical failures that can be adequately understood as fallacies only because they are misused tactics of reasoned dialogue. These are pragmatic failures of argumentation rather than semantic failures or errors of reasoning.

Another basis for erroneous interpretations and inferences lies in the expression "seems valid." This phrase can be taken to mean that there is a psychological element in the concept of a fallacy. But such an inference is misleading. In order for something to be a fallacy, to whom does it have to seem valid? This question continues to be the subject of some controversy in informal logic, despite the clarifications of Hamblin (1970) and van Eemeren and Grootendorst (1984).

Some exponents of informal logic insist that this discipline should study only actual (real-life, natural) arguments, as opposed to made-up or invented examples of arguments. Therefore, the conclusion is drawn, an argument is not suitable as subject matter for informal logic unless it was actually propounded (presumably seriously) by some real person. But such a criterion is absurd. If a particular argument was propounded by Bob Smith on Thursday, November 3, 1988, and was recorded and witnessed by his wife, Alicia, does that elevate it from the status of an "artificial example" to a real-life argument, suitable for study by informal logicians? The suggestion leads to trivial disputes about what a fallacy is. But anyone who argues that the concept of fallacy contains the element of deception or illusion had better be prepared to be accused of a simplistic psychologism of this sort.

Govier (1982) put forward a definition of the concept of a fallacy that included the notion of deception: "A fallacy is a mistake in reasoning, a mistake which occurs with some frequency in real arguments and which is quite characteristically deceptive" (2). In her explanation of this definition, Govier made it quite clear that she did not mean to claim that an arguer must intend to commit a fallacy (or have guilty intentions or deceptive motives of any sort) in order for us to say correctly that his argument was fallacious or that he committed a fallacy. She added (2) that a fallacious argument is one that "will seem like a good argument to many people much of the time." It is quite clear from Govier's article that she is expressly trying to avoid a crude psychologism in her definition of fallacy.

Yet that charge was expressly brought forward against her in a subsequent issue of *Informal Logic*. Carroll (1983, 23) criticized Govier's definition of fallacy by writing that she wrongly "considers *deception* to be an essential element of a fallacy." In her defense, Govier (1987, 200) felt compelled to respond that this criticism failed to distinguish between "actual deception" and "deceptiveness as a tendency to deceive" and to repeat her contention that actual deception is not required. Thus Govier had to put up an argument in order not to be impaled on the usual dilemma of psychologism.

In order to pin down a particular instance of an argument as fallacious, it should not be necessary for a critic to prove that the proponent of the argument had guilty motives or an intent to deceive. Nor should it be necessary for the critic to prove that the argument seemed valid to the person to whom it was directed or to any other individual. A fallacy should be a fallacy not by reason of any actual person's psychological beliefs that the argument seemed valid, invalid, attractive, repulsive, or believable. Rightly, according to Hamblin (1970; 1971), van Eemeren and Grootendorst (1984; 1987), and Walton (1984; 1987), a fallacy should be evaluated in relation to a normative model of dialogue that reflects the commitments of the proponent and respondent as evaluated by interpretation and analysis of the given text of discourse and the context of dialogue of a particular case.

What it means to say that an argument was fallacious because it was deceptive, or gave the illusion of being correct instead of the reality, relates to the use of that argument as a tactic in a context of dialogue. An argument used in a quarrel could be correct or appropriate, meaning that it contributes to the goal of the quarrel. Yet the same argument could be inappropriate or incorrect in a critical discussion. The deception or illusion of correctness can be explained through the dialectical shift. In the given case, the argument was used in an inappropriate context of dialogue.

In short the concept of "illusion" or "seeming to be what it is not" is important to the concept of fallacy because it is what makes informal logic an applied discipline. But the most facile, obvious, and usual ways of interpreting the expression psychologistically as 'seems valid but is not' tend to trivialize this pragmatic aspect of the concept of fallacy. This has perverted the idea of fallacy and has led to its denigration by logic textbooks.

The historical and etymological derivations of the word 'fallacy' outlined in the next section rightly suggest that a fallacy is more than an error of reasoning based on an illusion. It is a systematic type of attacking argumentation technique or tactic used in order to try to defeat an opponent by exploiting a weak point, laying a trap, or taking advantage of deception. This account of the concept of fallacy is quite consistent with Aristotle's study of the kinds of sophistical refutations in *De sophisticis elenchis*, where fallacies are treated as kinds of tactics that can be used to get the best of an opponent in argumentation in a verbal exchange or dialogue framework. This whole idea is a radical paradigm shift, however, from the point of view of modern logic.

The Greeks distinguished between what they called *eristic dialogue*, or purely contentious argumentation—a type of discussion exchange that is a "fight to the death"—and *dialectic*, which is a constructive type of dialogue that has a genuine capability of revealing insight or knowledge into a topic where there is a legitimate difference of opinion. According to Aristotle (*De sophisticis elenchis* 171 b 27), those who are bent on victory at all costs and stop at nothing to win, even unfair tactics, are contentious and quarrelsome arguers. A fallacy in this framework can be regarded as a misapplication of argumentation themes appropriate for a quarrel to another type of dialogue.

When we use the term 'persuasion dialogue' or the equivalent term used by van Eemeren and Grootendorst, 'critical discussion,' we refer to a normative model of reasoned dialogue where the goal of the enterprise is for each party to convince the other party that his point of view is right by means of certain "rational" types of arguments, that is, by using argumentation schemes appropriately. By the very nature of this type of dialogue, the goal of one party is opposed to the goal of the other, and therefore the flavor of the dialogue is somewhat adversarial—such dialogue therefore properly has a tactical aspect.

But the critical discussion is very different in its goals and methods from the outright quarrel, which is an all-out fight to defeat an opponent in dialogue with virtually no holds barred. The critical discussion has rules of discussion that are appropriate for each of its four stages. These rules guarantee that a genuine persuasion dialogue is, at least to some degree, a reasoned form of exchange that is qualitatively different in nature from the purely combative quarrel.

4. A Dilemma for Fallacy Theory

If the Aristotelian tradition or viewpoint of looking at fallacies as sophistical refutations is to be revived or used as the basis for a new fallacy theory, our concept of fallacy needs to be carefully rethought. It is no longer adequate to call a fallacy an argument that "seems valid but is not." This slogan is based on a shallow deductivist and psychologistic view of fallacy that is not useful or constructive in the project of building up a new fallacy theory. It is especially worth being aware of one danger.

It is important for fallacy theory to avoid being impaled on the horns of a dilemma that is posed at the current state of development of this newly emerging field. The dilemma is posed by two opposed conceptions of fallacy. One conception is the deliberate sophism—the intentional perpetration of a deceitful trick or fraud in argumentation by a perpetrator on a victim. The other conception could be called the paralogism—the failure of an argument to be valid because it fails to fit some structural (characteristically semantic) relation that the premises should bear to the conclusion. For example in the *Dictionary of Philosophy* (Runes 1964), we find a *paralogism* defined as "a fallacious syllogism; an error in reasoning" (225). And a *sophism* is defined as an "eristic or contentious syllogism; distinguished from paralogism by the intent to deceive" (295). The paralogism conception typically takes the form of the saying "A fallacy is an argument that seems valid but is not." The dilemma is posed by the presumption that we seem to be faced with an exclusive choice between these two competing conceptions of fallacy—we must define 'fallacy' one way or the other, it seems.

The danger here is that we might identify the sophistical tactics conception with the deliberate sophism conception in just the following way—we might require that any sophistical tactics fallacy must be a deliberate sophism, that is, it must be a deliberate deception by some actual person. The accompanying danger is that we might identify the error-of-reasoning type of fallacy with the paralogism conception, requiring that a fallacy is an error-of-reasoning type of fallacy only if it is an argument that seems valid to some actual person but is not.

Both of these identifications would be quite obstructive (fallacious?) for fallacy theory. Yet both of them are actually encouraged by the way the term 'fallacy' has come to be used in the traditions expressed in the logic textbooks. Ordinary English usage expresses a somewhat different meaning of the term 'fallacy.'

In English, as we noted above in section 1, the term 'fallacy' has become stretched too thin to include all kinds of errors, lapses, and mistaken beliefs. The book *Popular Fallacies* (Ackermann 1970) lists hundreds of popular or folk beliefs that were current at one time or another but have now been shown to be false by historical or scientific evidence. Included are items already noted like, 'A man has one rib fewer than a woman.' or 'Lightning describes a zigzag path with acute angles.' In this usage, the word 'fallacy' is taken much more broadly than its specialized use in logic to refer to failures of logical reasoning or misuses of argumentation tactics. The broad usage refers not (necessarily) to failures of logical inference but to any false belief that is or was popular. The term 'fallacy' must be defined more narrowly for use in logic.

Although it has generally been presumed within logic that this broad usage of 'false belief' is not appropriate for the logician's meaning of the term 'fallacy,' still the logic textbooks have taken too broad an approach, including all kinds of errors and weak arguments under the heading of fallacies. As Hamblin (1970) showed, this broad approach is out of synchronization with the original Aristotelian conception of fallacy as the use of a sophistical refutation in argument, a deliberate tactic of deceptive argument used unfairly to get the best of a partner in dialogue.

Interestingly, the origin of the word 'fallacy' clearly links fallacy with the use of crafty tactics used to deceive a coparticipant in argument. According to Klein (1971, 272), the English word 'fallacy' comes from the Latin *fallacia*, meaning "deceit, artifice, stratagem." According to Lewis and Short (1969, 721), *fallacia* means "deceit, trick, artifice, stratagem, craft, intrigue," and the associated verb *fallo* means "deceive, trick, dupe, cheat, disappoint." Going back further, the Greek verb *sphal* means to "cause to fall," as used in wrestling tactics, but the same word was also used to refer to verbal tactics, meaning "to cause to fall by argument." According to Paul van der Laan, the Greek term is descended from a Sanscrit term *sphul* (or *sphal*), meaning "to waver." These etymological origins of 'fallacy' clearly express the centrality of the idea of a fallacy as a tricky argument tactic used to trip up, or craftily get the best of, a partner in a verbal exchange, causing him to "waver" and "fall."

These older and more fundamental notions of fallacy were systematically built into a science of logic as a method for dealing with fallacies by Aristotle in the *Topics, De sophisticis elenchis*, and the *Rhetoric*, as we saw above. In the *De sophisticis elenchis* (171 b 22), Aristotle in fact compared sophistical argumentation to the use of unfair tactics in an athletic contest, describing contentious reasoning as a kind of unfair fighting in argument (see section 10 below). This part of Aristotle's logic, however, pretty well passed into oblivion for the next two thousand years, and the science of logic, after Aristotle, became syllogistic logic and subsequently propositional and quantifier logic. The term 'fallacy' evolved away from its root meaning.

The modern idea of fallacy prevalent in so many logic textbooks sees a fallacy as an argument that fails to be valid, for example, an invalid syllogism, but annexes a gesture to the root idea by adding that it is an argument that seems to be valid. The *Oxford English Dictionary* (1970, vol. 4, 45) lists, in addition to the meaning of sophism or misleading argument, a more current technical meaning: "In Logic, *esp.* a flaw, material or formal, which vitiates a syllogism." The *OED* notes, however, that this is not the meaning of fallacy used by Wilson's *The Rule of Reason* (1552), where deceit was included as part of the concept.

German, like Dutch, has two separate words to refer to the kinds of failures covered by the single English word 'fallacy.' According to *Harrap's Standard German and English Dictionary* (Jones 1967, 25), a *Fehlschluss* is an incorrect, wrong conclusion or a "wrong inference." In logic, this term means "fallacy or paralogism," but outside logic, it means "unsuccessful shot" or "bad shot" (25). According to a native speaker, this word naturally refers to failures of correct argument or inference where no intentional deception of one party by another is (necessarily) involved.

According to *Duden* (1981, 2637), *Trugschluss* is used in logic to refer to the use of deceit (*Täuschung*) or trickery (*Überlistung*) by one partner in dialogue (*Gesprächpartner*) to make the other draw the wrong conclusion. The aspect of intentional deceit involved in a *Trugschluss* is made explicit in the *Brockhaus* entry (1974, 49), where the word "intentional" (*absichtlich*) deceit is used. According to this entry, a Trugschluss is a type of Fehlschluss where the wrong conclusion is made to be drawn by one partner in dialogue, through the use of trickery or deceit. See the further explanation given in chapter 8, section 10.

The tradition of treating fallacies in the logic textbooks and manuals reflected this duality as well. Most of them tried to portray the concept of a fallacy according to the Fehlschluss model of an error of reasoning or argument that seems valid but is not (not surprisingly, without much success, or any notable improvement on Aristotle's treatment). Some, however—notably Bentham, Schopenhauer, and a few modern authors like Thouless (1930) and Fearnside and Holther (1959)—treated the fallacies, in practice, as tricky tactics of deception by a dialogue partner. But nobody seemed to notice, including these authors, that their treatment of the fallacies was a deviation from the "seems valid but is not" type of Fehlschluss conception of fallacy or an espousal of the sophistical tactics conception. Apparently the subject of fallacies struggled on in a very practical ad hoc way as a discipline without there being any serious or sustained interest in the theoretical question of what a fallacy is. The only exception would appear to have been the series of books on informal logic by Alfred Sidgwick, pretty well ignored by everyone (even Hamblin). Sidgwick was at least aware of the problem, although he did not appear to have proposed any solution to it.

The problem with the Trugschluss or concept of deliberate deception is that it requires that, in each case where we want to prove that a fallacy has been committed, we are required to establish that the perpetrator had a "guilty mind," an intent to deceive. This requirement would make the evaluation of fallacies a heavily psychological task. In fact, too psychological—it would entail the unfavorable kind of psychologism that Hamblin (1970) warned about. In advocating the normative model of dialogue as the structural device to aid in the determination of fallacies, Hamblin took commitment as the central idea, and he emphasized (264) that a commitment is not necessarily a belief of the participant in the dialogue who has it. According to Hamblin, we do not believe everything we say, but our saying it commits us to it subsequently, whether we believe it or not. Hence the evidence for or against a participant's having committed a fallacy is to be sought in the text of discourse given in the case and in the context of dialogue. This given text and context reflect the arguer's commitments, and that is what is important for us as critics when we decide whether or not a fallacy was committed in a particular case.

Van Eemeren and Grootendorst (1984) reaffirmed the importance of Hamblin's way of approaching the concept of fallacy when they warned (6) that it is necessary to guard against the internalization of the subject of critical argumentation by avoiding "psychologizing." One of the main features stressed by their approach to argumentation is *externalization*, the concentration on the expressed opinions of a participant in a discussion, and on the statements made by that participant in the discussion, as opposed to the "thoughts, ideas and motives which may underlie them" (6). Very much in the spirit of Hamblin's approach, the theory of van Eemeren and Grootendorst would appear to steer us away from the direction of thinking of a fallacy as an intentional deceit, at least where bad motive is an essential part of the concept of fallacy.

The task of "nailing down" a fallacy, then, should stop somewhere short of having to prove deliberate intent to deceive. What needs to be proved is that a particular technique of argumentation was used improperly or incorrectly in a given case, in a way that did not meet the requirements for the use of that type of argument in the context of dialogue for the case. Thus a fallacy is the abuse of a technique of argumentation in dialogue in such a way that the rules of procedure for that type of dialogue have been violated. But showing that such a violation has been committed in a particular case, while it does involve the commitments of the participants in the dialogue, should not require showing the existence of deliberate deception by one participant.

The problem with the second conception of fallacy is, first and foremost, that it stands no chance of doing justice to the analysis and evaluation of the major informal fallacies—especially those fallacies that can be classified under the heading of sophistical tactics.

There is something in the seeming-validity idea, but if this idea is construed in a simple, psychologistic way, it becomes a severe obstacle to the development of fallacy theory as a branch of logic, or the normative analysis of conversational discourse. Fallacies are, it is important to emphasize, powerful and effective techniques of argument generally. They are practically useful to study and guard against because they are the kind of argumentation tactics that often tend to work for strategic or deceitful purposes. Thus they are based on calculated tactics that work to trick people, and they are pitfalls that in fact do trip people and can fool us quite effectively. It should not follow, however, that every instance of fallacy has to be an intentional deceit perpetrated by a guilty proponent, or an argument that seems valid to a gullible respondent. This is itself a kind of fallacious argument that could perhaps be called the psychologistic fallacy.

What meaning of the term 'fallacy' is best, at least as an initial target for explication, for use as a technical term (or term of art) for use in logic and discourse analysis? Here it will be argued that the univocal concept of fallacy as approximately expressed by the use of the single term, as in English, is best generally preserved, but that the duality expressed, for example in German, points to the fundamental type of classification between the two main types of fallacies. We argue for a classification of fallacies into two basic types corresponding to the distinction between Fehlschluss and Trugschluss—errors of reasoning and sophistical tactics fallacies (sophisms).

It will be argued here, however, that the Fehlschluss-Trugschluss type of distinction is in some important respects too radical for good fallacy theory and that if it is preserved as it stands, such a classification would be a serious obstacle to the development of good fallacy theory.

Nevertheless, the originating root concept of a fallacy as the use of a verbal tactic of argument to cause an opponent in dialogue to fall or trip up is one that we need to return to.

5. Sophistical Tactics

Douglas Ehninger once compared a critical discussion to a finely tuned violin.[6] It must have the right balance of tautness and slack. Similarly, the persuasion dialogue has an antagonistic aspect. The stronger argument wins, and each side must try to build up the strongest argument for his side. In addition, each side must be prepared to be tolerant and open and to empathize with the point of view of the other.

The critical discussion is partly adversarial, and therefore argumentation tactics are important. But collaborative rules are also very important. Excessive or inappropriate quarreling is not only obstructive, a kind of fault in persuasion dialogue but can even be fallacious in some forms of unfairly aggressive tactics. These tactics might be not inappropriate in a quarrel, but they can be fallacious techniques of argumentation when used in a critical discussion.

It is vital to distinguish between the goal of a type of dialogue generally and the individual goals of the participants engaged in that type of dialogue. The goal of a critical discussion, for example, is to resolve a conflict of opinions by reasoned argumentation. But the goal of each participant is to convince the other party of the truth of one's own thesis (point of view) by reasoned argumentation. But how does one do this? Basically, what each participant needs to do is to prove his/her thesis from premises that the other participant is committed to. But how does one do this? Basically, one uses argumentation tactics.

Argumentation tactics are argumentation schemes coupled with argumentation themes, as used in a particular case by one participant in a dialogue to carry out his individual goal in relation to the situation of the other participant in the dialogue.

According to our new theory in section 6 below, a fallacy is defined as a type of a tactic or ploy of argument used inappropriately in a context of dialogue. A fallacy is more than simply a violation of a rule of reasonable dialogue; it is a deceptive tactic or trick of argumentation based on an illusion created by an underlying dialectical shift from one type of dialogue to another. There are always two parties to an argument containing a fallacy—the perpetrator and the intended victim. According to the *Latin Dictionary* of Lewis and Short (1969, 721), as noted in section 4 above, *fallacia* meant "deceit, trick, artifice, stratagem, craft, or intrigue." *Fallacia*, as we noted, comes from the Greek word *sphal*, meaning "cause to fall." This word was used by Homer to refer to wrestling, but it was also used in a more abstract sense of "cause to fall by argument," which refers to the use of verbal tactics of defeat in argumentation, as expressed by the new theory. As noted above, and in section 9 below, this is exactly reminiscent of how Aristotle explicitly compared contentious reasoning to unfair fighting in athletic contest.

Aside from Aristotle, however, there have been occasional—if sporadic and isolated—attempts to view fallacies as argumentation tactics. The first serious modern attempt to devise a list of common tactics to deceive an opponent in argument was the short work of Bentham (1969) on political fallacies. Bentham defined a fallacy as an argument employed "for the purpose, or with a probability, of producing the effect of deception" in another person with whom one is engaged in argument. Although confined to political examples, Bentham's list of tactics is generally interesting in its own right.

Schopenhauer (1951) was even more systematic and more general, offering a list of thirty-eight "stratagems" that can be used to get the best of an opponent in argument. Schopenhauer actually defined dialectic (11) as "the art of getting the best of it in a dispute," casting aside the Aristotelian presumption that dialectic has something to contribute to the discovery of truth. His aim was exclusively practical.

Thouless (1930), in an appendix to his book *Straight and Crooked Thinking*, presented a list of "thirty-four dishonest tricks" (249–58) that can be used "for detecting dishonest modes of thought" commonly encountered in arguments and speeches. Although largely practical, Thouless's treatment also has the normative aim of exposing these tricks of argument as "dishonest" tactics.

These various attempts to portray fallacies as argumentation tactics were not very successful and did not have much, if any, impact on mainstream developments in logic. The problem was that they lacked any coherent, underlying theoretical or normative basis that would enable a user to evaluate a given case as fallacious or not. It is certainly useful to know about such tactics, but from the viewpoint of logic, our goal is to evaluate arguments as correct or incorrect, not just to get the best of an argument.

Like the approach taken by these authors, however, the study of methods best serving the new theory should be inherently practical. A fallacy is portrayed by the new theory not just as a violation of a rule of dialogue but as the use of a technique that can be skillfully deployed in the real cut and thrust of argumentation. Learning how the fallacies work is a practical skill that is as much a matter of experience as of following rules (and perhaps more). Instructing someone on how to identify, confront, or deal with the fallacies is not a job that starts from scratch. A beginning level of skill or competence must already be presupposed, and it is a question of refining and improving these given skills. The abilities to understand an argument, identify missing premises, detect a conclusion, and so forth must already be presupposed to some extent. It is a question of enhancing an expertise by improving skills that already exist.

In the new theory, the normative and the practical are combined. A fallacy is the use of a tricky tactic but one that can be evaluated as inappropriate or incorrect in relation to a normative model of dialogue.

The study of fallacies has a normative element, a thematic element, and a practical element. To be a fallacy in the sense advocated in our theory, something must contain all three of these elements.

1. Normative Element. A fallacy is a serious violation of a rule (or rules) of reasonable dialogue.

2. Thematic Element. A fallacy contains a sequence of moves in a smooth pattern. It is a technique of argumentation that you need to recognize and become familiar with. You are most likely familiar with it already and are often using it in everyday argumentation. But by studying it at a somewhat higher level of abstraction, you can become more expert at dealing with it.

3. Practical Element. A fallacy is an effective device that can be used to make an arguer think he has received a convincing or suc-

cessful argument. It actually works, in practice, although it is of course not always effective, whenever it is used. It has to be applied to the right kind of situation in a given case. A fallacy is worth being warned about, and coached on, in preparation for encountering it in the everyday practice of argumentation. With preparation, you can learn how to cope with it better. Otherwise, it can take you by surprise and can more easily be the instrument of your defeat in a contested argument.

In short, argument tactics can have a positive side, in addition to their negative side, which is so often emphasized. Fallacies are powerful types of techniques that have been inappropriately used in a given case to defeat an opponent unfairly in argument. The notion of a device of argumentation being *used* in a certain way is very important here, however. The underlying argumentation scheme or theme used for the purpose of sophistical refutation of an adversary in a fallacious argument could also possibly be used for the purpose of constructing a proper, reasoned argument in a dialogue. What this point brings out is that fallacies have a strong pragmatic element—they are schemes of argumentation that are used in a certain way in a particular context of dialogue. Whether they are used appropriately or wrongly, for deceitful or constructive purposes, depends on the nature of the framework of the dialogue in which they are used.

The kinds of argumentation patterns cited in the current textbooks under the headings of the various informal fallacies are, in the preponderance of cases, types of attacks and defenses in two-person dialogue. They become comprehensible as important objects of study within a conception of argument as multiple-person dialogue where each participant has the aim of carrying out some goal of argument, like persuasion, in relation to another participant.

This statement was very true of the fallacies studied in chapter 7. In order to evaluate an instance of the ad baculum fallacy, you had to see it used as a particular type of tactic, a threat, used to influence another party in a dialogue who can be presumed to see it as a threat even if the threat is expressed covertly. Or in the case of many questions, you have to analyze the profile of dialogue as a bilateral sequence of exchanges, where the commitments of the respondent, and the proponent's use of his knowledge about them, are crucial to the fallacy.

The sophistical tactics type of fallacy is defined as the use of an argumentation tactic in a given situation where there is a presumption that the participants are supposed to be engaged in some type of dialogue. The fallacy arises through the grafting of this abstract

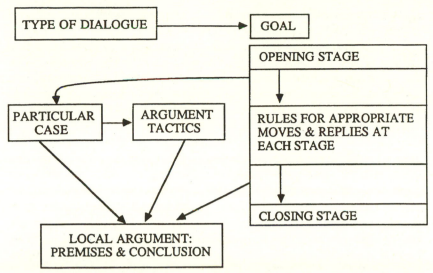

Figure 10   Components of a Sophistical Tactics Fallacy

framework of dialogue onto the particulars of the given case. This sort of structure is called a *frame* in artificial intelligence—a general framework that has loose ends of porous surfaces that fit onto the particulars of a given set of circumstances. Out of this "fitting onto" arises the use of argument tactics—techniques used to fulfill the goals of an abstract model (a normative model) of dialogue by fitting it to the given information in a particular case, as indicated in figure 10. The tactics are the means for carrying out the goals in a specific situation, for example, a dispute about a particular controversial topic where the participants can be divided into two sides, each side with a given position and point of view that is laid out with given particular details (many of which are typically incomplete and unknown).

A fallacy of a particular type is not identified with the violation of a particular rule of a particular type of dialogue—or, at any rate, not that alone. It is a type of argument scheme or theme that has been misused in a particular way in relation to the goals and rules of a type of dialogue that the participants in the argument are supposed to be engaged in. It is a kind of argument scheme or theme that could be used correctly or appropriately to contribute positively to the goal of a type of dialogue. But its use in the given case is fallacious because it has been turned to other ends—it has been used to subvert or obstruct the proper goals of the type of dialogue the participants were originally supposed to be engaged in. Throughout, the sequence of argumentation has been used as a tactic that goes against the purpose of coherent and constructive type of dialogue.

Typically, however, a fallacious argument looks plausible because there is a dialectical shift inherent in the case. The use of a particular type of tactic is really not appropriate or constructive for the type of dialogue the arguers are really supposed to be engaged in. Because it would be appropriate in some other type of dialogue, however—which does in fact seem to be partly involved or appropriate in the particular case as given—the argument does have an aura of prima facie reasonableness or correctness. For example, the ad baculum derives this aura of seeming acceptability from its use in a context of negotiation or quarreling.

Typically, then, a fallacy is an argument tactic that would certainly be appropriate in some context of dialogue, but that context is not the one that properly fits the given case. Such an argument looks not entirely implausible, because it would fit some context of dialogue. The question of evaluation is one of whether it fits the given case. If not, it is a fallacy.

This concept of fallacy is inherently pragmatic, because the underlying question always to be asked in evaluating a particular case is: what is the context of dialogue? Whether the argument is or is not fallacious, according to this approach, always depends on what the purpose of the discussion is supposed to be.

In general, the new theory implies that much more is involved in the concept of fallacy than rule violation and that fallacy is a practical concept essentially tied to skill in the use of argumentation techniques in a range of cases similar to, but also different from, each other. In teaching students to recognize and cope with fallacies, then, according to the new theory, the job is one of building on and enhancing human expertise, in both the use and the analysis of argumentation tactics in use in everyday conversations.

6. The New Definition of Fallacy

The new theory of fallacy presented below is built around the functional theory of argument in Walton (1992c, chap. 5), where a key distinction is made between reasoning and argument. *Reasoning* is defined as a sequence of propositions (premises and conclusions) joined into steps of inference by warranted inferences. *Argument* is then defined as a use of reasoning to contribute to a talk exchange or conversation called a dialogue. So conceived, reasoning is a narrower notion that is compatible with the point of view of traditional logic, whereas argument is a frankly pragmatic notion that has to do with the uses of reasoning in a context of dialogue.

The concept of a fallacy has an inherently dual nature or an ambiguity implicit within it. On the one hand, there is a tendency to see a fallacy as being a faulty inference from premises to a conclusion—an error of reasoning. This tendency is most marked in traditional treatments with respect to the formal fallacies. On the other hand, there is a tendency to see a fallacy as being a deceptive trick, a misleading argument used to get the best of a partner in a speech exchange or dialogue. So conceived, the fault is more than just one of incorrect reasoning but is a sophistical refutation (to use the Aristotelian term), a use of argument to deceive a partner in dialogue. Traditionally, however, this second conception of fallacy has been incoherent or undefinable to logicians because there was no pragmatic framework in which it could be defined or could make coherent sense.

Now that we have (in chapter 4) given a pragmatic basis of the different frameworks or types of dialogue in which argumentation occurs, however, we can explicate this second conception of fallacy as a type of failure of the correct use of an argument in a context of dialogue.

The definition of the concept of a fallacy given below is pragmatic in the sense that a fallacy is more than just a faulty inference from a set of premises to a conclusion. Although the definition will include this aspect, it also goes beyond it, defining a fallacy as a type of failure of the Gricean cooperative principle (CP). The CP of Grice (1975, 67) states that any contribution to a conversation must be "such as is required, at the stage at which it occurs, by the accepted purpose or direction of the talk exchange." With fallacies, we are dealing with arguments that fall short of this requirement, according to the new theory advanced here. A fallacy is defined as an argument that not only does not contribute to the goal of a dialogue but actually blocks or impedes the realization of that purpose. A fallacy is defined as a purported argument that goes counter to the direction of the talk exchange and poses a serious danger to blocking it.

The definition of the concept of a fallacy now proposed has five clauses, each of which is a necessary condition of something being a fallacy. A fallacy is (1) an argument (or at least something that purports to be an argument); (2) that falls short of some standard of correctness; (3) as used in a context of dialogue; (4) but that, for various reasons, has a semblance of correctness about it in context; and (5) poses a serious obstacle to the realization of the goal of a dialogue. Each clause of this proposed definition is highly controversial and must be argued for by reference to the analysis of specific fallacies.

Even the first clause is controversial, for it is often questionable in the traditional treatments whether certain "fallacies" are arguments. For example, it is often said or implied in the textbooks that ambiguity is a fallacy, whether the ambiguity is in the form of an argument or not. Also, equivocation is a fallacy because what appears to be a single argument is, in reality, many arguments. To deal with this type of case, our approach is to relax the definition to require only that a fallacy must be any move in a dialogue that is supposed to be an argument.

Another case in point is the fallacy of many questions and other question-asking fallacies. For a question is not (on the face of it) an argument. Another problem area is the *ad* fallacies, where questions can be raised about whether a threat, personal attack, or appeal to pity is a fallacy even if it is not (explicitly) in the form of an argument, with premises and a conclusion. Hence much of the work of establishing this definition of the concept of a fallacy involves addressing substantive questions of how best to analyze the individual fallacies (especially the major informal fallacies outlined in chapter 2). Chapter 9 of this book will be partly occupied with carrying on this task.

The definition, in the form given above, is rather a long one, and it would be good also to be able to encapsulate the central idea in a shorter form, slogan, or quick definition. This we do as follows: a *fallacy* is a deceptively bad argument that impedes the progress of a dialogue. This short form is, of course, only a slogan that sums up the longer form of definition, each part of which needs to be clarified, justified, and qualified.

The purpose of offering a definition is to attempt to coordinate the field of informal logic in this area where there is widespread disagreement and uncertainty on how to identify and evaluate the various fallacies. Part of the problem is the uncertainty and unclarity of what is meant when it is said that such-and-such is a fallacy or that such-and-such an argument is fallacious, because of different points of view on what constitutes a fallacy (or, in some cases, even the absence of any clear standpoint). Another part of the problem is the inherent ambiguity of the notion of fallacy (noted above). The definition is univocal, but it needs to be seen how it can cope with the inherent ambiguity of the word 'fallacy' in English in a way that offers a helpful conceptual building block for informal logic.

All fallacies have an argument core of reasoning contained in them, and all have some degree of contextual involvement (use of argument in a context of dialogue). Some are more dialectical (contextual) than others, however. For this reason, there is an ambiguity inherent in the concept of fallacy. Two subtypes of fallacy need to be distinguished.

Paralogisms are errors of reasoning that relate to logical forms of inference. These forms can be deductive or inductive forms of argument like modus ponens or arguing from a sample to a larger population. Or they can be argumentation schemes for presumptive reasoning. Paralogisms are fallacies that arise chiefly through failure of an argument to meet a set burden of proof.

Sophisms are dialectical fallacies that relate to a use of argumentation in a context of dialogue. They are extended sequences of argumentation whose fallaciousness is revealed by examining a *profile of dialogue,* a connected sequence of dialogue moves that is an exchange of responses between two parties who are arguing with each other. Sophisms are bad arguments because the sequence of moves reveals a characteristic type of deceitful sophistical tactic that hinders the correct progress of a dialogue. Typically, sophisms seem correct and appropriate only because there has been a dialectical shift to a different type of dialogue from the one the participants were originally supposed to be engaged in.

By contrast, paralogisms seem correct because of the apparent use of an argumentation scheme or a form of reasoning that is (in principle) correct. The formal fallacies outlined in chapter 3 are the classic examples of paralogisms.

7. Properties of the New Concept of Fallacy

There are six characteristic properties of the new concept of fallacy. The new approach is:

1. Dialectical. The main normative model is that of a two-person exchange of moves in a sequence of argumentation. Whether an argument is fallacious depends on the stage of a dialogue that the arguer is in.

2. Pragmatic. The context of dialogue is extremely important in determining whether a fallacy has been committed. You (as critic) must interpret and analyze the text of discourse (extended sequence of discourse) of the particular case.

3. Commitment-Based. The arguer's commitment at a given stage of a dialogue is a key concept in determining whether a fallacy has been committed. This acceptance-based approach does not, however, rule out or denigrate the role of deductive, inductive, or knowledge-based reasoning.

4. Presumptive. Fallible (defeasible, default) reasoning is very important to understanding how the fallacies work in everyday discussions. The major fallacies involve weak, fallible kinds of argumentation, like argument from authority, argument from sign, and so forth, that are successful if they shift a burden of proof in dialogue.

5. Pluralistic. Several models of dialogue are involved. The critical discussion is important, but it is not the only type of dialogue in which argumentation occurs. The notion of a dialectical shift is the key to understanding how fallacies work as arguments that seem correct.

6. Functional. The concept of the use of argumentation themes in argumentation is very important. A fallacy is more than just a rule violation of a type of dialogue—it has to be seen as a particular technique of argumentation that is used inappropriately by one party in a dialogue against another party.

Fallacies are techniques of argumentation that have been used in a counterproductive way to steer a discussion away from its proper goals or even in an aggressive attempt to close off the effective possibilities of an adversary's critical questioning in the dialogue. But identifying the pragmatic context of dialogue is the key to fixing the claim that an argument is fallacious. An aggressive personal attack that could be perfectly appropriate for an outright quarrel, as an effective tactic to hit out verbally at your opponent, could be highly destructive to the balance required for fair and constructive persuasion dialogue (critical discussion). In that context, the use of the same technique of argumentation could be shown to be a fallacy. In a scientific inquiry, yet another context of dialogue, the same use of the technique of personal attack could be even more outrageous and clearly out of place. In this context, it could even more easily be shown to be a fallacy, by showing how the tactic used is inappropriate as an acceptable method of working toward the goals of the dialogue.

According to the definition proposed by Johnson (1987, 246), a fallacy is "an argument which violates one of the criteria/standards of good argument and which occurs with sufficient frequency in discourse to warrant being baptized." This definition is very favorable. It eliminates the need for psychologism by focusing on frequently used types of arguments that are "baptized." But the new pragmatic theory goes beyond Johnson's definition in two ways.

First, baptism, according to the new theory, should be defined in relation to general types of arguments or techniques of argumentation that are worth labeling, studying, and watching out for because they are dangerous, that is, relatively powerful, as well as common in argumentative discourse. In deciding which argumentation techniques to baptize, we should be guided not just by "perceived frequency," as advocated by Johnson (247), but also by the extent to which the technique, used in key situations, is able to swing a weight of presumption. Johnson is right, however, to conceive of baptization as a pragmatic device and to note that the student "must learn to dispense with the label" in some cases.

Second, a careful distinction needs to be made between the general concept of fallacy as a type of argumentation technique and the concept of a fallacy as a particular instance of a fault or failure of argumentation in a given case. This distinction becomes especially important on the new pragmatic theory. For the type of argumentation technique typifying the fallacy can be used correctly in some particular cases. To tell whether an argument is fallacious in a particular case, then, according to the new theory, requires judgment. We must look and see how the technique was used in relation to the particulars of the given text of discourse and the surrounding context of dialogue.

The new theory is pragmatic because it involves a judgment of how well a technique has been used in a particular case. There are three characteristic trajectories of use. A satisfactory execution of a technique results in a correct argument (correct according to the standards of the type of dialogue). A weak execution of a technique results in an argument that is open to critical questioning. Such an argument may be said to be "weak," or to "have a fault" but not in a strong sense, meaning that it is fallacious. "Weak" means, in this sense, insufficient or incomplete. A misuse or abuse of an argumentation technique, in the new theory, is in a separate category from these first two trajectories of use. A misuse of an argumentation technique is a misoriented execution that is at odds with the context and purpose of the dialogue in question. This misuse of an argumentation technique results in a fallacy. It could be a tricky, deceptive use of a technique, deliberately designed to cheat an opponent in argumentation, or it could be an underlying, systematic error in the execution of the technique, without there (necessarily) being any intent to trick or deceive someone. It is this third category of use—wrong use of a technique—that results in the commission of a fallacy.

These three kinds of trajectories of misuse are summarized in figure 11. Notice that *weak execution* is included in the category of *right use*, not *wrong use*. In this theory of fallacy both the categories of *weak argument* and *fallacy* come under the category of *error* (*fault*) in argument use. But a fallacy is a very special and serious kind of error—not an intentional error or deliberate abuse of a technique, necessarily. Instead, it is defined as a misdirected execution—the use

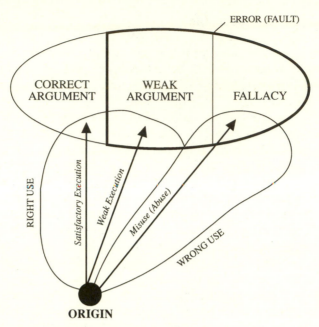

CORRECT ARGUMENT

WEAK ARGUMENT

FALLACY

RIGHT USE

Satisfactory Execution

Weak Execution

Misuse (Abuse)

WRONG USE

ORIGIN

Figure 11   Trajectories of Uses of Techniques

of a tactic to block or prevent legitimate goals of reasonable dialogue from being implemented.

There are certain practical techniques of argumentation that can be used in different contexts of dialogue but that are worth "baptization," or special classification and labeling, as powerful (and therefore potentially dangerous) techniques, as they are so often used in everyday argumentation practices. These include, for example, the following three techniques: (1) attacking an opposed argument by attacking the arguer's sincere willingness or cooperativeness in engaging in collaborative dialogue (ad hominem); (2) supporting your own side of an argument by citing expert opinion (ad verecundiam); (3) invoking presumption—declaring that your side of an argument must stand (prevail) because the other side has not given sufficient evidence to refute it (ad ignorantiam). All three of these techniques are inherently reasonable kinds of argumentation practices in the sense that they can be used appropriately, in some instances, to fulfill legitimate purposes of reasonable dialogue.

But they can also be used badly, wrongly, or inappropriately in some cases. Yet they are such powerful and common techniques that we need to be on guard, lest they be used against us by a contentious or unscrupulous arguer or even lest we carelessly or unthinkingly fall into using them uncritically or erroneously in our own thinking.

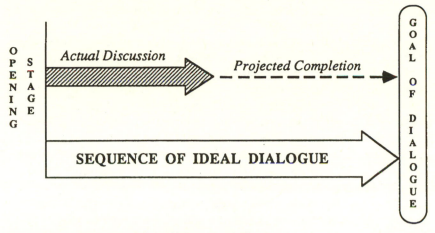

Figure 12 Evaluation of a Fallacy in a Dialogue

This respect for the danger implicit in the possibility of their misuse has led to a long-standing tradition of labeling them as "fallacies," meaning that they are always incorrect whenever they are used in argumentation.

But according to the new theory, argumentation techniques are not necessarily used incorrectly as "fallacies" whenever they have been employed in a particular case. Hence an evaluation of each particular case on its individual merits is required, as shown in figure 12.

The problem of evaluation is no longer one of "Fallacy—yes or no?" but one of asking more specific questions in a case. How far has the actual discussion moved toward the goal of the dialogue, as far as we can tell from the text of discourse of the discussion, as it has been completed? Is the actual discussion moving toward the goal? Is it falling short of the goal? Or is the discussion proceeding in such a wrong direction, due to some distortion, misdirection, or blockage, that it will never reach the goal if it keeps proceeding along these same lines? How bad is the problem? Is it a fallacy, or just a low level of argumentation that is not going along very well?

According to the new theory, the task of evaluation is reconceived as a tripartite classification. Is the particular case a correct argument, a weak argument (open to critical questioning), or a fallacy?

To understand this new pragmatic conception of fallacy, you need to appreciate that a fallacy is associated with the use of an argumentation scheme or theme in an orderly sequence of moves in interactive question-reply dialogue between two (or more) participants. To judge whether an argument is fallacious or not in a particular case, we must examine the particulars of the sequence critically insofar as the profile can be reconstructed from the given text and context of discourse. It is a question of how the technique of argumentation was used in that particular case. Was it used well or badly? How well or how badly it has been used are questions of interpretation, analysis, and evaluation that require judgment in assessing the wording of a particular dialogue exchange.

Basically, there are right uses and wrong uses of an argumentation technique. A *right use* is a use that supports the goals of the dialogue. But this can be strong or weak. A use of a technique that fulfills a proper goal of the dialogue, meeting the requirements of its argumentation scheme, is a *satisfactory execution* of that technique. A use of a technique that does not fulfill a proper goal of dialogue but nevertheless goes some way toward supporting its realization, is a *weak (partial) execution* of that technique. A *wrong use* of an argumentation technique is the turning of that technique toward some goal other than a proper goal of the dialogue in the case in question. This turning away from the goal is a kind of misuse (abuse) of an argumentation technique that is correctly associated with committing fallacies.

8. The Charge of Fallacy

The new theory of fallacy views the concept of fallacy in a dialectical framework. The allegation "Fallacy!" is a kind of charge put forward by one participant in reasonable dialogue against another participant. To be sustained, the charge must be backed up by evidence, or it fails to hold up. A charge of fallacy, therefore, carries with it a burden of proof for the proponent who has made the charge.

Substantiating a charge of fallacy, according to the new theory, is going to be a lot harder than the facile practices of the standard treatment presumed. This is bad news for those authors of textbooks who have been taking such a lighthearted approach to the study of fallacies in the past. But it is good news for the development of this field as a serious and mature area where scholarly research is possible. One of the main problems to be addressed by this developing field is that of pinning down a charge of fallacy.

Pinning down a charge of fallacy is a problem of context—it is a question of how the argumentation technique has been grafted onto the particular case. At the general level, you have a type of argumentation technique that is being used. And it is a question of the application of this general technique to the particular case, that is, the text and context of dialogue as these are extrapolated from the given text of discourse.

You need to identify the dialogue situation in the given case and then to understand how the general technique that was used fits into that situation. Then you can ask about rules of dialogue and determine which rules were allegedly broken by which party at which points in the sequence of dialogue.

The rule violation *by itself*, however, does not identify the kind of fallacy that is operative in the given case, nor does it show how that technique was used in relation to the tactical situation. For these reasons, a fallacy is not definable or understandable simply as a violation of a rule of reasonable dialogue. A rule violation must be involved, but understanding or evaluating a fallacy must also involve understanding how a type of argument was used in a context of dialogue. This means understanding how the argument in question was used to subvert or exploit the legitimate goals of the dialogue. The job is one of mapping the type of argument (the argument form, scheme, or theme) onto the particulars of the given case in a context of dialogue.

Analyzing an allegation of fallacy involves asking whichever questions are appropriate from the following list of eleven thematic questions.

1. What type of dialogue is involved? What are the goals of the dialogue?

2. What stage of the dialogue did the alleged failure occur in?

3. Could there be more than one context of dialogue involved? Was there a dialectical shift, at some point, that could affect the question of fallacy?

4. Was there a specific failure in an argumentation scheme alleged? What kind of failure or shortcoming was it? What type of argument was used? How was it deficient or used inappropriately?

5. How bad was the failure? Was it a fallacy or more just a blunder or weakness? Should it be open to challenge or refutation?

6. What maxims of dialogue were violated? Was failure of relevance involved?

7. Was the problem of sophism in an argumentation theme? What general technique of argumentation was used? How is the fallacy revealed in a profile of dialogue?

8. Who are the parties in the dialogue? What are their roles in the dialogue? How was burden of proof distributed?

9. How was the technique used by the one party against the other party as a tactic using deception (as indicated by the answer to question 3)?

10. Was the technique used as a calculated tactic of deception by its proponent? How aggressive was the use of the technique? How persistent?

11. Which critical questions were not answered? Or was critical questioning diverted or shut off by the use of the technique? Was there a chance to reply?

In any particular case, a glib answer to the question of whether or not a fallacy has been committed is no longer possible. A lot of work needs to be done identifying the context of dialogue and assessing the information given in the particular text of discourse. New methods will have to evolve—methods of practical logic and methods of discourse analysis.

What the new pragmatic theory tells us is that there are two levels involved in answering these eleven thematic questions in a particular case. At the general level, we need to understand each of the fallacies as techniques used as argumentation tactics in different contexts of dialogue. At the practical level, we need to see how these techniques can be employed properly or improperly in particular cases. This practical work has an empirical aspect—through the analysis of paradigm case studies, we can see how the techniques are employed in different ways, according to the requirements of special dialogue situations.

At the dialectical level, the counterpart of the weak argument is the *flawed argument*, which is a missing step or a gap in the sequence of moves required in order to carry out a successful sequence of moves in a dialogue. A flawed argument is typically an argument that goes some way toward its objective but leaves out certain key steps or requirements. The key difference between a fallacy and a flaw in an argument is this. In the fallacy type of case, there is an underlying systematic pattern of argument strategy that has been used in a way that goes strongly against the legitimate goals of dialogue in the given case. In the flawed argument, however, the main thrust and direction of the argument is consistent with the rules and aims of the dialogue, but gaps, missing parts, or questionable junctures make the argument fall short of its objective. Sometimes it is hard to prove whether a sequence of moves is fallacious or merely flawed. And sometimes it does not matter greatly, provided the flaw is noticed and understood as something that is a critical failure.

So far then, fallacies may be contrasted with flaws, blunders, and other weaker or less dramatic failures of argumentation, because a fallacy involves the use of a characteristic pattern of strategy of argument in discussion in order to extract some advantage or win out over an opponent in a contestive discussion unfairly. But this is not the only respect in which a fallacy is a distinctively strong form of tactical misuse of an argumentation technique.

When something is called a fallacy in the new theory, it means not only that the argument so labeled was part of a strategy of attack but also that the attack has gone pretty badly wrong. To say that an argument is fallacious or commits a fallacy is therefore quite a strong form of censure implying that the use of argumentation technique underlying the argument must or should be wholly rejected because it is based on some sort of underlying, systematic fault or deficiency. But of course we most often criticize arguments without going quite this far.

A *criticism* of an argument could be defined as a challenge to the argument that questions a weakness of the argument by citing a specific shortcoming. Here the term 'shortcoming' is a generic term for any kind of blunder, flaw, error, or weakness of execution. At the extreme end, some criticisms could allege that the shortcoming is so serious that a fallacy may be said to have been committed. There can, however, be criticisms of arguments that do not necessarily allege or show that the argument has committed a fallacy. By contrast, a fallacy could be characterized as a serious type of weakness, deficiency, breach, or misuse of an otherwise reasonable type of procedure in an argument or move of argument, open to criticism to the extent that the argument can justifiably be judged to be strongly refuted. A fallacious (particular) argument can then be defined as an instance of argument where a critic can show, by appeal to reasonable guidelines of dialogue, in relation to the information given in this particular case that the argument commits a fallacy.

What have been called "fallacies" by the textbooks are—more soberly construed in a dialectical perspective—often criticisms that are reasonable in some cases and not so reasonable in others. The 'not so reasonable' category actually ranges over many kinds of blunders, flaws, and deficiencies of argument. Only in the more extreme and severe cases of this sort is the label 'fallacy' justified. And a criticism always has two sides. From the point of view of the critic it is a good argument, one that at least poses a significant critical question for the other side to answer. From the point of view of the arguer criticized, however, a criticism is something to be defended against by offering rebuttals, if possible, or at least explanations or clarifications.

For example, the circumstantial ad hominem argument (chapter 5) is a form of criticism that questions the consistency of an arguer's commitments, citing a presumptive contradiction between the arguer's personal circumstances and his argument. This kind of criticism typically does not refute the argument criticized, but it does raise questions about the sincerity or integrity of an arguer's advocacy of his own argument. This sort of criticism, as shown in chapters 5 and 7, raises critical questions for both the attacker and the defender.

9. The Balancing Aspect of Argumentation

The key to understanding what a fallacy is lies in the idea of presumption. Generally, the way to win an argument is to get or preserve the weight of presumption on your side of the argument. In all critical discussions on controversial subjects, conclusive evidence, to determine which side in the controversy is right, is inevitably insufficient. The outcome is therefore determined on a basis of whether the balance of presumption can be tilted toward one side or the other. Now, there are right ways to shift a balance of presumption, but there are also powerful and effective ways of forcing the weight of presumption against your opponent's side of a disputed argument that are fundamentally not right (not justifiable). These powerful tactics are abuses of right ways of shifting a burden of presumption, and they fall into certain categories or commonly used patterns. It is these commonly employed patterns of illegitimate presumption shifting that are the underlying structures of the informal fallacies.

Hence it most often turns out that the informal fallacies are, at bottom, revealed as moves in argumentation that are strongly open to censure as serious errors of reasoning because they are argumentation tactics used in an aggressive effort to win out over an opponent in an argument by preventing that opponent from continuing with the argument. Indeed, in many cases, they function as kinds of tactics designed to close off the line of argument altogether or to prevent the opponent from arguing at all or from taking a real part in the argument. A critical discussion always has an open quality (open-mindedness). And the real fallaciousness of an informal fallacy often characteristically turns out to be the forced and premature application of techniques aiming for closure of the dialogue by the aggressive tactical maneuver of one side. Thus a fallacy is often a kind of tool of argumentation used as a technique for sealing off the line of further dialogue, shutting an opponent up.

Not all attempts to occupy the high ground by unfairly or unjustifiably trying to swing the weight of presumption to one's own side, and against one's opponent's side, of an argument are, however, fallacious. Informal fallacies are certain types of tactics of aggression in argumentation that tend to fall into powerful and commonly used patterns that can be identified and studied.

Although the term 'fallacy,' as currently used in logic textbooks, contains significant distortions and exaggerations of meaning that make retention of this usage problematic, the term is worth retaining in certain cases, provided it is appropriately redefined.

Now that we have seen how three of the traditional so-called fallacies work, it will indicate the general nature of the reform needed. The argumentum ad hominem, appeal to expert opinion, and the argumentum ad ignorantiam have been revealed as reasonable arguments (in some instances)—that is, as instances of argumentation schemes that have a proper and important role to play in reasonable dialogue. But we have seen how they can also be used (in other cases) as fallacies.

To understand the concept of fallacy, it will be necessary to see how there are distinctive types of techniques used in argumentation to shift a burden of proof properly. Such techniques can be used as offensive or defensive tactics. When held properly in check, in a balanced and restrained way, these tactics can make a valuable contribution to a reasoned discussion by testing the strength of an argument or the defense of an argument. When unleashed in an unfairly aggressive and purely contentious manner, however, such tactics can undermine or destroy the balance needed to sustain a fruitful critical discussion.

In dialogue on a contentious issue, the thesis of the Proponent is opposed to the thesis of the Respondent. If the issue is truly a contentious one, that is, is open to contention by the disputants, then neither thesis is known to be true, or false, at the outset of the dialogue. That is, each thesis is open to being proved or disproved by the arguments of the participants in the dialogue. At the beginning of the dialogue, then, there is a certain balance between the thesis of the Proponent and the thesis of the Respondent. Any line of argument that serves to make the thesis of one side more plausible will ipso facto make the thesis on the other side less plausible, because the two theses are the two sides of a balance. Whatever makes one side go up must make the other side come down, and conversely. This balance idea seems to be related to the concept of burden of proof. To fulfill the burden of proving your own thesis is automatically to bring forward evidence that refutes your opponent's thesis. In other words, any relevant argument will tilt the burden of proof one way or the other, thus affecting both theses at the same time.

Any argument that disturbs the balance necessarily affects both sides. It follows that any "positive" argument is therefore also, ipso facto, a "negative" argument. Any argument that tends to confirm one's own thesis to be proved must also serve to refute or undermine the other side's thesis.

This balancing aspect of argumentative dialogue is present not only in the strongly opposed type of dialogue (the dispute), where the thesis of each participant is the opposite (negation) of the thesis of the other. It is also essential to the weakly opposed type of dispute, where one party is questioning or doubting the thesis of the other party. The reason is that any good argument that increases doubt will ipso facto decrease the plausibility of the thesis defended by the other side by opening it to further questioning.

The reason why the balancing effect is characteristic of dialogue on contentious issues generally is that the issue in this type of dialogue is *open to contention*. Because firm commitment to one side or the other is open at the outset of the dialogue, the balance of evidence could swing either way. To transfer commitment to one side is always to lighten commitment to the other side. This seems to be a pervasive feature of persuasion dialogues generally, because persuasion dialogues are always on questions of values (conflicts of opinion), which originate in a *stasis*, a kind of problematic suspension of decision about which side is right on an issue. Moreover, the same kind of balance characteristic also seems to apply to negotiation dialogues, and to many other types of dialogue as well, because all of them start out and are based upon a conflict or difference of point of view between the two participants.[7]

If this reasoning is right, it may be possible to explain why there has been a prevalence of emphasis on the negative and adversarial in the traditional literature on argumentative dialogue—an emphasis on fallacies, refutations, trapping the opponent in contradictions, and the like. This emphasis on apparently negative and unfriendly aspects of argumentation does not stem from some vengeful desire to embarrass people or put them down by proving them wrong. Instead, in matters of unsettled opinion, commitment is inherently unfixed—or should be, if we are not dogmatic about such things—and therefore commitment on an issue will tend to settle on the side that is least weak, least open to objections and doubts. Because, characteristically, positive evidence and knowledge to support one's thesis on a question of opinions and values tend to be lacking or inconclusive, the best or only way open to support the thesis is by questioning or finding weak points in the alternative or opposing theses on the same issue. Hence the characteristic type of argument tactic is to attack the opposing points of view.

This adversarial aspect of dialogue often makes it seem like a weakness of the method of dialogue as a form of reasoning, because the exponent of dialogue seems to be favoring negative criticism and personal attack over the positive approach of collecting knowledge, for example, in the form of scientific evidence and experimental confirmation of a hypothesis. But on the other hand, we might say that the alleged weakness is really a kind of strength, because persuasion dialogue can apply to matters of doubt and uncertainty where resolution of the problem by scientific verification has not worked, or has not been possible, for various reasons. The burden of proof is a device that allows a persuasion dialogue to come to a provisional conclusion that nevertheless settles an issue, instead of merely discussing a controversial issue endlessly.

The problem remains of showing precisely how a critical discussion can achieve this goal of yielding truth or insight. A clue is that this advance toward yielding insight into an issue comes through a mutual refinement of each other's viewpoints by the participants, with the result that their viewpoints are refined or clarified by the testing of dialogue.

Typically, fallacies occur when one party to a critical discussion adopts a quarrelsome attitude, trying to get the best of the other party by using sophistical tactics or misdirected reasoning in an inappropriately aggressive way that is suitable for quarreling but is a poor, inefficient, and unfriendly way of contributing to the critical discussion. The problem is not just in the use of a particular argumentation tactic per se. It is the inappropriate use of it—or indeed the abuse of it—in the context of dialogue (the critical discussion) that the participants were supposed to be engaged in. The problem with fallacies is that the use of such tactics actually blocks or interferes with the goals of critical discussion.

10. The Dilemma for Fallacy Theory Revisited

Any attempt to construct an analysis of fallacy appears to be impaled, as we saw in section 4 above, on the horns of a dilemma. For the analysis, it appears, must choose between the following two alternatives. Is a fallacy always an intentional deception, where one participant in an argument is trying deliberately to mislead the other participant? Or is a fallacy a kind of mistaken or erroneous inference that can occur without any deliberate deception where one party is trying to trick another into accepting a bad argument? The second kind of "fallacy" could occur where a wrong conclusion has been mistakenly drawn—even in solitary deliberation, for example—without its being a case where one party is trying to deceive or trick another party in an argumentative exchange between the two.

This dilemma is a genuine problem. Some fallacies, like post hoc, hasty generalization, formal fallacies, and composition and division, seem to fit more naturally into the second category, while others, like ad verecundiam or ad baculum, for example, seem to fit more naturally into the first category. There is also evidence that we may distinguish two meanings of 'fallacy,' one of which is closer to the first category, and the other closer to the second category, as we saw in chapter 2.

The roots of the English word 'fallacy' are to be found in the Latin word *'fallacia,'* which is in turn etymologically connected to *falsus,* the Latin word for 'false.' It would be a mistake to infer from this genuine connection, however, that a fallacy can simply be any false statement, as far as logic is concerned. According to Lewis and Short's *Latin Dictionary* (see section 4 above), the word *falsus* can mean "false," or alternatively, it can mean "deceptive, pretended, feigned, deceitful, or spurious." What can be "false" about a logical fallacy is its deceptive use to cover up a logical weakness in an argument.

A deeper insight into the roots of English practices of usage comes from the original meaning of the Latin word *fallacia,* from which the English word 'fallacy' was etymologically derived. The roots of the word are important enough to clarify a bit further here, beyond the account already given in section 4 above. According to Klein's *A Comprehensive Etymological Dictionary of the English Language* (1971, 272), the English word 'fallacy' comes from the Latin *fallacia,* meaning "deceit, artifice, stratagem." Revealingly, according to Klein (272), *fallacy* is connected to the Latin verb *fallere* (to deceive), and *falsus,* the past participle of *fallere,* is the origin of the English word 'false.' According to Lewis and Short, *A Latin Dictionary* (1969, 721), *fallacia* means "deceit, trick, artifice, stratagem, craft, intrigue," and the verb *fallo* means "to deceive, trick, dupe, cheat, disappoint." Also (721), the adjective *falsus* means "deceptive, pretended, feigned, deceitful, spurious, false." In view of the current usage of the word 'fallacy,' this connection between 'fallacious' and 'false' is curious and revealing. Going back further is even more revealing, however.

The term 'fallacy,' as noted in section 4 above, was originally descended from a Sanscrit word *sphul* (or *sphal*), which means 'to waver.' This term is the root of the Greek word *sphal,* which has two meanings. First, *sphal* means 'to cause to fall' and was used (e.g., by Homer) in this sense to refer to tricks and strategies of wrestling. Second, *sphal* can also mean to deceive or trick in a more metaphorical and verbal sense, which could be expressed as "cause to fall by argument." The English word 'fallacy' is not directly from Greek, however, but is based on the Latin verb *fallere* (to deceive or trick), as described above. Even so, the root Greek idea is of the utmost importance.

The idea that the sophistical arguer, or sophist, is one who is skilled in clever techniques of getting the best of an adversary in argument, and who can teach these skills to his students, is one that was highly familiar in ancient times. In view of these ancient presumptions, it was not at all strange to compare the sophistical reasoner to a skilled athletic coach who can instruct an athlete on how to use tricks and stratagems to get the best of an opponent in an athletic competition. For Aristotle (1955), as noted many times already, the comparison was perfectly natural. "For just as unfairness in an athletic contest takes a definite form and is an unfair kind of fighting, so contentious reasoning is an unfair kind of fighting in argument; for in the former case those who are bent on victory at all costs stick at nothing, so too in the latter case do contentious arguers" (1955, 63). Aristotle clearly saw fallacies frankly as dialectical shifts, perceiving that a contentious argument may be appropriate in a quarrel but inappropriate and fallacious in a scientific context of reasoning like an inquiry in geometry. This dialectical viewpoint has, however, not been familiar to logicians after Aristotle and is only now beginning to gain something of a small foothold once more. Small wonder, then, that Aristotle's concept of fallacy has appeared incoherent and alien to modern readers.

It is possible once again to breathe life into these ancient ideas in a workable theory of fallacy, however, by basing it on a pragmatic and dialectical notion of the use of argumentation in a context of dialogue. This pragmatic notion views argumentation as a collaborative, goal-directed sequence of speech acts, for example, questions and replies, which take place within a global context of dialogue. Fallacies, like other failures of collaborative politeness in dialogue, can be evaluated through the application of normative models of reasonable dialogue—what Hamblin (1970) called formal dialogues—to particular cases, or what Hamblin called realistic dialogues.

Every dialogue has a goal, and the participants have strategies for trying to fulfill these goals according to the rules. But in applying a strategy to a particular case, the participants must use techniques of argumentation in a contributory manner. As applied to a given case, in adversarial dialogues these techniques can become *tactics*, particular partisan uses of the techniques for making a point against the opponent or attacking an opponent's point of view at some juncture in an argument. But tactics can be used unfairly in such cases. Whether an argument, in a particular case, is fallacious should depend not on an intent to deceive but on how the argument was used in that particular case, in relation to the maxims or accepted procedures, based on the Gricean CP, for that type of dialogue. What is important, then, is not the particular arguer's purpose but the purpose of the dialogue as a whole. Tactics that might be quite appropriate in a quarrel or negotiation, for example, might be fallacious if used in argumentation in a critical discussion. Fallacies are like techniques of street fighting that might be highly effective tactics to use against a mugger but would be quite out of place when used in a wrestling or boxing competition in the Olympic Games, for example.

The idea that a fallacy must contain an intent to deceive, or must be some sort of deliberate attempt to get the best of the other party in an argument unfairly, is based on a confusion between the common goal of dialogue in an argument and the individual goals of the participants in that dialogue. The occurrence of a fallacy in argumentation should not be equated (necessarily, or in a one-to-one correspondence) with the existence of an intent to deceive by one of the arguers. Such a naive and unsupportable kind of psychologism would make the critical evaluation of fallacies as failures to meet normative standards of correct argumentation unworkable. Instead, a fallacy should be defined as a technique of argumentation used in a way that strongly goes against the collective goal or purpose of a cooperative dialogue.

Of course, not all argumentative dialogue is fully cooperative. Such dialogue often has a strongly adversarial element. But all dialogue requires some sort of cooperation or collaborative following of basic procedural rules; otherwise, it would not really be dialogue. Even the quarrel has rules and requires a certain degree of cooperation.

The occurrence of a fallacy in a particular case of argumentation is not to be based on the intent of the participant. It should be identified with the misuse of an argumentation technique that goes against (hinders, blocks, prevents) the implementation of the (joint, global) goals of the dialogue that is the proper context, or normative background, of maxims of polite collaboration for that type of conversation.

If a Fehlschluss represents the error or blunder of argumentation, and a Trugschluss represents the intentional deception or trick of argumentation, a fallacy is neither of these things. It is somewhere in between. It is a misuse of an argumentation technique, used in a way that goes against goals of reasonable dialogue. It is often, and perhaps in the cases of some of the fallacies, paradigmatically used as a tactic of deception to get the best of another party in argumentation. But it can also be committed without realizing that one has committed it, even in solitary deliberation. A fallacy does not need to be intentionally committed, but it is more than just an error or blunder in argumentation. Although it will no doubt prove a subtle idea to grasp, the best concept of fallacy for logic has to exist in the middle area between these two extremes.

9

Putting the Theory to Work

Of the major informal fallacies described in chapter 2, five have already been analyzed as a basis for developing the theory of fallacy presented in chapter 8. In chapter 9, nine more of these fallacies are analyzed on the basis of the theory. This shows, at least to some extent, how the theory can be put to work as an aid to developing an applied logic of the fallacies that is practically useful.

Enough has already been done on the concept of relevance in chapter 6, as a basis for future investigations of the ignoratio elenchi fallacy, or fallacy of irrelevance. And the post hoc fallacy has already been given a dialectical analysis in Walton (1989a, 212–33) that the reader can be referred to. So in chapter 9, my commentary is confined to the following nine major fallacies so far untreated: ad ignorantiam, ad verecundiam, argument from consequences, slippery slope, equivocation, secundum quid, straw man, equivocation, amphiboly, and accent.

My goal is not to give a complete analysis of any of these fallacies. The current literature on them is in a varied state of development. On some, there is a growing body of work, while on others, little of any real use exists outside the textbooks. What I can best do here is to show how the new concept of fallacy places each of these individual fallacies in a broad framework that will be useful as a research program for moving ahead with the project of identifying, analyzing, and evaluating these fallacies in a productive and well-organized way.

While much useful work has already been done on some of these fallacies, one frustrating problem is that there has been little agreement, or useful body of results, on exactly what is meant when each of them, as a type of argumentation, is said to be a fallacy. This, then, is what we need to show—that the meaning of 'fallacy' worked out in chapter 8 can be applied helpfully, and with some promise of success, to the remaining major fallacies, on the basis of what is known about them now.

1. The *ad Ignorantiam* Fallacy

Arguing from ignorance, as we have seen in chapter 5, section 7, is by no means a fallacious or erroneous type of argument in every instance. In fact, all presumptive arguments could reasonably be described as species of arguments from ignorance. Such arguments should be criticized as erroneous or unduly weak arguments from ignorance only where the natural order of reasonable dialogue or inquiry required to meet the standard of proof has been waived or ignored and a premature presumption has been accepted or promoted. The context of dialogue for ad ignorantiam is often that of either the critical discussion or the inquiry. In either case, a particular ad ignorantiam question or argument should be evaluated in relation to the *openness* or *closure* of the inquiry at the particular stage of its development relative to the particular corpus of argument being assessed. One needs to ask: is the closed-world assumption applicable or not? For example, in case 83, it was presumed that the railway schedule was epistemically closed or complete, that all stops were indicated. As a result we were able to conclude definitely that if a stop at Schipol is not marked on the schedule, then the train does not stop at Schipol.

But then again, the question in many cases is a matter of degree or weight of presumption. In general, the basic critical question to be asked in analyzing a particular case is this: how far along has the process of dialogue gone? The initial stage is the formulation of a problem or the posing of an allegation. Here the argument is open. Once closure is declared or agreed upon, the process of open argumentation is terminated. Hence the stage of dialogue is the critical thing.

The evaluation of an argument from ignorance also depends on the type of dialogue the argument was advanced in. In an inquiry, the burden of proof tends to be very high, as opposed to say, a critical discussion. In an inquiry, the argument from ignorance takes the following general form as a subspecies of the argumentation scheme of the argument from ignorance.

A has not been established as true (false)

Therefore, **A** is established as false (true)

There are four critical questions appropriate in this situation.

1. What stage of the inquiry are we in?
2. What (if anything) counts as "being established" at this stage of the inquiry, that is, what is the burden of proof (if one is appropriate at that stage)?
3. Has the burden of proof demonstrably not been met at this stage?
4. Can this failure to meet the burden of proof rightly be taken to imply closure of the inquiry?

The use of the term 'established' in the argumentation scheme above suggests the inquiry as the appropriate context of dialogue and suggests that closure of the inquiry is at issue. Therefore, the danger of weak argumentation, blunders, or other kinds of error arises through the failure to clarify the real stage that the dialogue is in by considering the four critical questions above.

On the other hand, as was consistently emphasized in chapter 5, section 7, the argumentum ad ignorantiam basically reflects the idea of burden of proof, which is a legitimate and important part of all reasoned, interactive argumentation. Hence it is to be emphasized that there is nothing inherently wrong with the argumentation scheme above per se.

The idea that burden of proof is a legitimate, even essential part of all reasoned dialogue is brought out even more dramatically when the context is that of a critical discussion or a deliberation, where a high burden of proof may not be appropriate. In these cases, the argumentum ad ignorantiam is based on presumption, and the subspecies of the scheme can be represented as follows.

A is not presumed to be true (false)

Therefore, **A** is presumed to be false (true)

The four critical questions for this situation parallel those for the one previously presented.

1. What stage of the dialogue are we in?
2. What is the burden of proof at this stage?
3. Has the burden of proof not been met at this stage?

4. Does the burden of proof and the strength of the opposing arguments at this stage license a shift of the sort described by the argumentation scheme?

The critical questions above reflect the idea that a shift in the burden (obligation) to back up a presumption can vary at different stages of a dialogue. All presumptive argumentation is like a balance, where even a small shift of the burden of proof to one side or the other can radically affect the outcome of the argument (see chapter 8, section 9). Everything depends on what stage of the argument we are in and which side has the burden at that stage.

The difference between the first and second variant argumentation schemes above corresponds to an important distinction cited by Woods and Walton (1982, 120). A thesis is said to be *refuted in the strong sense* if it is shown to be false at the outcome (closure) of a dialectical exchange. A thesis is said to be *refuted in the weak sense* if the discussion shows that the respondent has clearly insufficient grounds for holding that thesis at that stage of the discussion. The danger of fallacies and other errors comes in when there is potential for confusion between the strong and weak senses of 'refutation.'

When the situation is appropriate for weak refutation (the second type of argumentation above), the correct model to represent a particular case is that of a weak presumption shift. It may be a perfectly reasonable and appropriate argument, provided that all the critical questions can be answered reasonably. When the argumentation is a strong refutation (comparable to the first type of argumentation above), however, there is greater danger of logical mischief afoot because this stronger type of argumentation tends to be utilized when moving toward closure, especially in a type of dialogue like the inquiry, which has a strong burden of proof. The danger here should alert us to Locke's warnings (Hamblin 1970, 159–160) about the danger of pressing ahead too hard to prevail against an adversary (see section 2).

In general, it is a good thing to remember here that any argumentative dialogue can be broken down into the four phases outlined in chapter 2. In the first phase, the *opening stage*, the issue and procedural rules of the argument are set out, and obligations and burdens of proof are laid down. In the *confrontation phase*, the nature of the obligation (burden of proof) of each party is defined or agreed upon. In the *argumentation phase*, the argument is carried out, and evidence and arguments both for and against the contention in dispute are put forth, challenged, and evaluated. In the fourth stage, the *closure stage*, the argument comes to a close, and the issue is resolved or finally evaluated. Once closure is declared, decided, or agreed upon, the evidence is regarded as complete unless the argument is reopened.

In commonplace cases of argumentation, precise junctures of these three phases may not be clearly indicated. In some more clearly regulated contexts, however, like a board of directors meeting or a parliamentary debate, regulations governing the precise limits of each of these three phases may be decisively enforced.

A good example of how the stages of an argument evolve is the sequence of inquiry characteristic of the criminal law trial procedure. This procedure of inquiry breaks down into a sequence of stages.

1. Accusation: Is the allegation serious or well founded enough to justify carrying it to court?
2. Trial Set: This stage already requires some evidence against the defendant.
3. Trial: Arguments for both sides are set out by the questioning of witnesses and other participants.
4. Verdict: A conclusion is reached on the basis of the evidence presented in the trial.
5. Appeal: The process of inquiry may be reopened in exceptional cases. But this step requires *new* evidence, that is, evidence not previously considered in stage 3.

The criminal trial is essentially a persuasion type of argument, but it has strong elements of the inquiry mixed in as well at certain points. Although scientific experts are often called in to testify, the judge or jury must arrive at a decision based on plausible reasoning for the most part. As nonexperts they must try to deduce plausible conclusions from what the experts say, interpreted through the cross-examination dialogue of the attorneys. Other areas of law are based more on the bargaining model of dialogue. And in fact, many persuasion dialogues in criminal law can now shift away from the issue of guilt as the discussion turns into a speech event of plea bargaining. This is a shift from persuasion dialogue to the negotiation type of dialogue.

In the opening phases of an argument, an agenda may be set that defines the issues, the participants may discuss the rules to be followed or other procedures that they can agree on, and generally the difference of opinion or interest to be resolved will be articulated. In the middle phase, the process of debate or inquiry is undertaken.

The error of ad ignorantiam, or erroneous argument from ignorance, occurs where the inquirer leaps too quickly ahead and arrives at a dogmatic conclusion without going through the steps of inquiry or dialogue required to establish the conclusion queried. The failure is a failure to meet the burden of proof required for the context of dialogue, because the arguer has not answered the critical questions that clarify the burden of proof appropriate for the stage of dialogue. But an error of this kind can be either a fallacy or a blunder. The ad ignorantiam fallacy occurs where the arguer adopts tactics of trying to close the argument in his favor by suppressing the critical questioning of the other side or by trying preemptively to thwart these questions.

The classic case is the use of bare innuendo and suspicion in case 13. Here we know from the context that the investigation was little more than a "witch hunt" to label political enemies as "Communist sympathizers." The ad ignorantiam is here used as a smear tactic to suggest by innuendo that a person may be presumed guilty on the grounds that no evidence is available or has been brought forward by him to show that he is innocent of the charge. Such a reversal of burden of proof is used here as a tactic to block a proper investigation or trial from taking place by pressing ahead with the accusation in an obstructive way.

There is an identification problem with the ad ignorantiam argument, because lack of knowledge in argumentation is generally partial rather than total. For example, in case 83, someone might say that this is not an argument from ignorance, because we know (positively) that the schedule says that the train has no stop at Schipol. But this is really a quibble, because it is somewhat arbitrary whether knowledge is described in positive or negative terms. For example, if I look on the coatrack and see clearly that there are no coats on it, is this positive knowledge of the fact that the rack is empty, or is it a kind of negative knowledge that there are no coats on the rack, as far as I can see? Such "positive" and "negative" knowledge is generally mixed in argumentation. Hence all arguments from ignorance are better described as partially arguments from knowledge and partially arguments from lack of knowledge. Even so, it is not generally difficult to recognize arguments from ignorance. Once one starts recognizing arguments from ignorance, one begins to see that they are a lot more common than one might initially have thought.

2. *Ad Verecundiam* as a Fallacy

The expression *argumentum ad verecundiam* (appeal to respect, reverence, or modesty) should not be used as a generic term for arguments that appeal to authority or to the authority of expertise. This expression should be used to refer to fallacious appeals to authority in argument—that is, to the abuse of appeals to alleged authority or expert opinion in order to prevent a respondent in argument from asking critical questions in further dialogue on an issue. Moreover, such fallacious instances of the argumentum ad verecundiam are to be distinguished from instances of critically weak, faulty, or incomplete appeals to authority in argument, where the proponent has failed to back up one or more of the required steps of documentation corresponding to the three premises of the argument scheme for the argument from expert opinion, given in chapter 5, section 7.

It is particularly important to distinguish between two primary meanings of the term 'authority' in this connection—the cognitive authority and the administrative authority. Cognitive authority is the authority of expertise, based on the knowledge, judgment, and advanced skills in a field, characteristic of the expert practitioner's practical reasoning and theoretical knowledge in that field. Problems of concern in informal logic come to the fore when this expert knowledge is extracted in a clumsy manner from the source expert or used in argumentation in a way that is not consistent with the goal of a dialogue.

In case 16, the lecturer's implausible claim should be open to critical questioning by the audience, but Herbert's reasonable and appropriate questioning is cut off and dismissed by Helen on the grounds that Herbert is not an expert nutritionist himself. This tactic is a heavy-handed, fallacious ad verecundiam used forcefully to block off the proper flow of critical questioning in a dismissive manner.

The argumentation scheme for the argument from expert opinion, and set of matching critical questions for it given in section 3 of chapter 5, show how specific shortcomings to document premises of this scheme can result in weak and inadequately supported appeals to expert opinion in argument. But a weak argument is not necessarily a fallacious argument. Moreover, numerous indications have shown (Walton 1989a, chap. 7) that where the argumentum ad verecundiam is fallacious, the problem is due to the misuse or dialectically inappropriate deployment of the technique of appeal to expert authority in argumentation in contexts of dialogue that can vary.

These indications point to the usefulness of a *dialectical analysis* of the fallacy of argumentum ad verecundiam that occurs in a particular case when a proponent, who cites an expert opinion in order to persuade a respondent in dialogue of some conclusion, wields appeal to authority in too strong a manner in order to prevent the respondent from replying with critical questions on the expert opinion. The fallacy here has to do with how the argument is presented and not just with the premises (propositions) that support the conclusion representing the expert's cited opinion.

In case 15, the proponent used an appeal to an expert in biology (Darwin) to support a claim in a different field (morality), thus leaving herself open to an inability to answer critical question 3 for the argumentation scheme of the argument from expert opinion. Is this a fallacy or merely a blunder? The more information we have from the context of dialogue in the case that the proponent tries to stick dogmatically to this dubious claim or to brush it aside by aggressively attacking the respondent as a nonexpert, the more evidence we have that it is a fallacy and not merely an error or weak use of argumentation from expert opinion.

An insightful kind of dialectical analysis of the fallaciousness of the argumentum ad verecundiam can be found in the brief but interesting comment in Locke's *Essay* (published in 1690), quoted by Hamblin (1970, 159–60). Locke describes the appeal to the opinion of an authority with a reputation of learning as a kind of argument "that men, in their reasonings with others, do ordinarily make use of to prevail on their assent." He does not see this kind of argument as intrinsically unreasonable or fallacious, but he does see it as inferior to "arguments and light arising from the nature of things themselves," which he calls argumentum ad judicium. So considered, then, Locke sees the appeal to authority in argumentation as acceptable in some cases, but he sees the argumentum ad judicium as preferable when it is available.

Already, Locke has alluded to the dialectical nature of the use of appeal to authority in argument when he described it as a kind of argument that one person uses in reasoning with another person in order to prevail on the other's assent. But he goes even further, explaining how this process of "prevailing upon another's assent" can be carried too far, where one party attempts to appeal to the pronouncement of an authority "to awe" the other party "to silence their opposition." What Locke describes, then, is an attempt by an aggressive participant in argumentative dialogue to use the awe or respect of the other for an authority to browbeat the other into submission.

The whole passage where Locke contrasts the argumentum ad judicium with the ad verecundiam, ad hominem, and ad ignorantiam types of argument in dialogue is relatively short and self-contained. The reader is referred to Hamblin (1970, 159–60) for a quotation of the whole passage, and to chapter 10, section 5 below, but the part specifically on the argumentum ad verecundiam quoted below (159–60) is especially interesting.

The first [type of element] is to allege the opinions of men whose parts, learning, eminency, power, or some other cause has gained a name and settled their reputation in the common esteem with some kind of authority. When men are established in any kind of dignity, it is thought a breach of modesty for others to derogate any way from it, and question the authority of men who are in possession of it. This is apt to be censured as carrying with it too much of pride, when a man does not readily yield to the determination of approved authors which is wont to be received with respect and submission by others; and it is looked upon as insolence for a man to set up and adhere to his own opinion against that of some learned doctor or otherwise approved writer. Whoever backs his tenets with such authorities thinks he ought thereby to carry the cause, and is ready to style it impudence in anyone who shall stand out against them. This I think may be called *argumentum ad verecundiam*.

Locke begins by giving several different reasons why a particular individual's opinion may have the reputation and esteem to be considered an authority, including "learning, eminency, power, or some other cause." Thus he would appear to be including administrative authority as well as cognitive authority. Certainly he is including political authority as well as expertise in a domain of knowledge comprising academic or scholarly fields of learning. He appears to include more popular types of opinion leaders as well when he refers to "reputation in the common esteem" and "approved writer" other than a "learned doctor." But he does not say that there is anything fallacious, or inherently unreasonable, in any of these kinds of appeal to authoritative opinion.

The fault of argument Locke does cite is the deployment of authority in dialogue by investing it with dignity in such a way that any attempt to question the appeal appears to be insolent or impudent. Here, then, is Locke's explanation of the phrase *argumentum ad verecundiam* as meaning "appeal to reverence or modesty." To question the opinion of a dignified authority who has been set up as a spokesman with an impeccable reputation backed by learning and eminence could appear to be a "breach of modesty" or a kind of "insolence." What Locke calls the argumentum ad verecundiam is the tactic of using appeal to authority in such a confident manner that the one who uses it poses his argument as so overwhelming and decisive that he is not prepared to tolerate any opposition. As Locke puts it, the perpetrator is "ready to style it impudence in anyone who shall stand out against them." Thus for Locke, argumentum ad verecundiam can be used as a sophistical tactic—in our terms, we could call it a fallacy—the fallacy of overzealous wielding of appeal to authority in a tactic of prevailing on another individual's assent in dialogue argumentation. It is a specious tactic to forestall critical questioning by investing one's cited authority with an infallibility that the one to whom the argument is directed dares not challenge without (apparently) offending the standards of politeness of the discussion.

What is alleged to be fallacious, on this analysis, is not the reasoning in the appeal to authority itself, but the manner of presentation of the argument from authority in the context of dialogue between two participants in argument. The proper role of a respondent in persuasion dialogue is to ask critical questions at the next move, once an argument has been advanced by a proponent at any particular point in the sequence of dialogue. But the problem pointed out by Locke is that if the appeal to the opinion of an established authority is advanced by the proponent of the argument using this appeal in an overbearing manner, the respondent may have great difficulty performing his rightful function at the next move without appearing to be impolite. By deploying this tactic of browbeating in argument, the user of the argument from authority is preventing the respondent to his argument from performing his rightful function in the persuasion dialogue. Thus the fallacy of the argumentum ad verecundiam is the infelicitous use of the opinion of an authority as a technique of argumentation to subvert the procedures of reasoned dialogue, which should allow for the free questioning of arguments used by both sides in the dialogue.

Locke's remarks suggest an emphasis not so much on the specific propositions in which the expert opinion is delivered (the reasoning in the argument) as on the manner in which the argument is presented in the sequence of exchanges that has transpired between the proponent and respondent in a persuasion dialogue. The fault is not a paralogism but a sophism.

What is wrong is that instead of being presented (appropriately) as a defeasible, presumptive inference—which is open to critical questioning by its nature—the argument from expert opinion is presented as if it were a tight deductive inference that the respondent cannot question. The tactic is (unjustifiably) to present the argument from expert opinion as certain (beyond doubt) and deductively closed, thereby not allowing the respondent the necessary room to reply. Evidence to substantiate this charge in a given case is to be sought in the profile of dialogue, showing how critical questioning is managed by the arguer, who has appealed to expert opinion to back up his argumentation.

The argument from expertise, in itself, may be somewhat plausible. As an instance of reasoning, the argument may carry a legitimate weight of presumption. But that is not the problem of the fallacy. The problem of the fallacy lies in its manner of presentation as a technique used in the sequence of question-reply dialogue in a context where a proposition vouched for by an external "authoritative" source of knowledge is being used by one party to persuade another party to accept a conclusion. But what sort of dialectical profile is presupposed by this type of argument?

It is an inherently valuable feature of a critical discussion to allow for external sources of knowledge to be appealed to by the participants. The participants could agree, in advance of the argumentation phase of the dialogue, to accept as premises any propositions brought forward by either participant from a given knowledge base, say, an encyclopedia. Such a proposition would have a special standing in the argument, meaning that a proponent who had brought it forward could presume that the respondent would accept it, at least for the sake of argument, unless the respondent immediately challenged or questioned the proposition, at the next move. Accordingly, a proposition from the agreed-upon *authoritative source* would have a special standing as a plausible presumption.

A procedure of this general type, called an *intersubjective testing procedure* (ITP) in van Eemeren and Grootendorst (1984, 167), is a testing method that enables participants in a critical discussion to use new information from a source previously agreed upon by both of them—for example, the source might consist of encyclopedias, dictionaries, or other reference books. Using this device, the participants in a critical discussion can take advantage of mutually shared, presupposed background knowledge (166). Such propositions can be challenged or retracted in some instances, however, according to van Eemeren and Grootendorst (166): if disagreement arises in discussion about these implicitly accepted propositions, both parties can deny that they are committed to a particular proposition.

This approach shows how two parties in a critical discussion could take advantage of information drawn from an authoritative source accepted by both of them in advance, which would give propositions drawn from this source a certain privileged (but not sacrosanct or nonretractable) standing in an argument advanced by one of them. Hence a proposition drawn from an authority—like a third party accepted by both participants as an expert in the domain under discussion—could be brought forward in argument by one participant, on the expected assumption that the other participant would be inclined to accept it as plausible. By this approach, then, the appeal to authority in argument could have a dialectical justification—meaning that it could be advanced as a kind of prima facie plausible presumption in a context of dialogue.

In this dialectical framework, an argument citing an authoritative source in the way indicated above could be used wrongly in various ways. A source not previously accepted as authoritative by both parties could be appealed to. Such a move could be erroneous but not necessarily fallacious. But perhaps even worse, an ostensibly authoritative source could be presented or used in argument in such a manner that the respondent is given no chance to retract his commitment to the proposition in question or even to challenge it by asking for reasons to back it up. The problem here may be a matter not so much of the content or source of the proposition at issue as of how it has been presented as an argumentation theme in the critical discussion. It is the contention here that the ad verecundiam fallacy generally comes under the heading of the sophism type of fallacy, over and above the specific paralogisms inherent in the use of the argumentation scheme for the argument from expert opinion, for example, citing a source who is no expert at all or is not an expert in the appropriate field.

With the argumentum ad verecundiam, then, the argumentation scheme is fundamental in evaluating the appeal to expert opinion as a correct or incorrect, strong or weak instance of presumptive argumentation. But the most severe, misleading, and dangerous types of cases of the ad verecundiam fallacy are more than just violations of, or failures to meet, the requirements of the argumentation scheme for the argument from expert opinion. They also involve a profile of dialogue giving evidence of use of a systematic tactic of browbeating the other party in a dialogue by pressing ahead too aggressively with the argument in a way that actually impedes the goal of the dialogue they are engaged in.

3. Argumentation from Consequences

Argumentation from consequences is a very common type of reasoning used in everyday conversation, especially in deliberation, planning, and advice-giving dialogue. It is a presumptive type of reasoning concerned with hypothetical conjectures about what will, may, or might happen in the future. It typically becomes fallacious when it is used by a proponent to try to intimidate the respondent by a kind of innuendo suggesting that the bad consequences are very scary and that the future is very uncertain or dangerous, without backing such claims up adequately. Hence the fallacious use of argument from consequences is often closely related to the tactics of intimidation used in the ad baculum fallacy.

Provided it is used in a way that conforms to the requirements of the argumentation scheme in chapter 5, section 9, however, the argument from consequences is a presumptively reasonable (nonfallacious) kind of reasoning that is used to shift a burden of proof in a dialogue.

The main type of fallacy that tends to occur with argumentation from consequences is based on a dialectical shift. Typically, two parties are having a critical discussion based on a conflict of opinions, and the one party argues to the other: "If you persist in holding that opinion, such-and-such bad consequences will [may, might] happen to you." The shift here is from a critical discussion concerning whether an opinion is right or not to a kind of advice-giving dialogue where the one party uses practical reasoning to warn the other of bad consequences of holding this opinion or making it known that he holds it. Of course, any such warning could also be interpreted, in some cases, as a covert threat. Hence the connection here with the ad baculum fallacy.[1]

Case 34 is a case of this sort. The two politicians are having a critical discussion on the issue of whether a woman should have the right to an abortion. The one politician argues for the so-called prolife side, but then the other one replies: "If you take that view, you will not be elected." Now, this could be interpreted as a licit dialectical shift, where the second politician is making a practical advice-giving aside that is not meant to have a bearing on the critical discussion about abortion. If so, there would be no fallacy. But the reason that we perceive case 34 as an instance of fallacious argumentation from consequences is that we see the reply of the second politician as an illicit intrusion into the critical discussion on abortion to use the practical consequences of the first politician's holding his particular point of view as a reason for claiming that this point of view is false (or cannot be defended as a right opinion in the critical discussion). This is a fallacy because it involves an illicit dialectical shift from the one type of dialogue to the other. Thus fallacious argumentation from consequences is best analyzed as a dialectical fallacy.[2] It is based on the use of a particular argumentation scheme, but the basic fallacy is of the dialectical shift type.

The dialectical shift in such a case can also be described as a failure of relevance. The second politician's reply can be described as not relevant dialectically in the sense that the practical warning makes no real contribution to the critical discussion on whether a woman should have the right to an abortion. Sometimes, too, the practical claim is weakly supported or implausible as well as being irrelevant. In case 33, the claim that questioning the justice of the U.S. side in the Mexican War of 1848 "would give comfort to our enemies" simply does not seem very plausible. This claim in itself, however, is not fallacious. It is simply a failure to answer adequately the first critical question of the argumentation scheme for the argument from consequences (chapter 5, section 8). The fallacy resides in the failure of relevance of this claim, even if it were true or well supported. Similarly with case 32—the problem is one of relevance.

4. The Slippery Slope Fallacy

Slippery slope arguments are of the four types identified by their characteristic argumentation schemes in chapter 5, section 10—the causal, the precedent, the sorites, and the full slippery slope argument. All these types of argument are, in principle, presumptively reasonable. Mainly they become fallacious when pressed far beyond the weight of presumption they can reasonably bear. The causal slippery slope argument is a special type of argumentation from consequences.[3]

There is something of an identification problem with the slippery slope fallacy. The four subtypes of slope argument are all different from each other in some respects.[4] One might well ask, then, what is it exactly that they share in common as slippery slope arguments? And also, what exactly is the difference between the causal slippery slope argument, as a type of slippery slope argument, and negative argumentation from consequences, which is (presumably) not a slippery slope type of argument?

The difference is that in a slippery slope argument, the proponent is using a repeatable sequence to warn the respondent that if he takes a first step, this sequence of subsequent steps will be embarked upon by the respondent in such a way that there will be no turning back until the horrible ultimate outcome occurs. The analogy to illustrate this characteristic is between a staircase and a slide. At any point partway down the staircase, you can stop and go back up the stairs again if you choose. On a slide, however, once you have taken that first step where your motion forward starts, there is no turning back.

The key difference between the two types of argumentation is in the retractability of the respondent's commitment. In the slippery slope type of argument, once you have committed yourself to that first step, then you cannot retract your commitment to each subsequent step because of the repeatable sequence of the chain of steps. Commitment to the whole thing is assured by the sequence, which makes the whole thing nonretractable, once a commitment (the first step) has been made.

The same notion of nonretractability of commitment is present in negative argumentation from consequences, but it is a simple one-step argument that does not essentially require a sequence. Argumentation from consequences can be chained together in sequences, but such a chain of argumentation would not be a slippery slope argument unless the sequence is of the repeatable and gradualistic type characteristic of the slippery slope argument.

Each of the four subtypes of slippery slope argument has this characteristic of pushing forward nonretractably along the sequence, once a first step or series of steps has been taken. With the sorites slope, it is the vagueness of a term, creating a gray area where no exact line can be drawn to stop the argumentation pushing ahead. The sorites slope is often combined with the precedent type of slope argument, which works by argumentation from analogy between pairs of similar cases. The causal slope argument works because of the repeatability of a causal sequence, for example, contagion of a disease or addiction to a drug. The full slippery slope typically works because of public acceptance—once the public becomes accustomed to a certain practice or privilege, it is natural to push things a step further. But in all four subtypes, the characteristic of the repeatable sequence and its structuring of nonretractability of the respondent's commitments is essentially the same, in how it functions as a type of argumentation.

The slippery slope argument becomes a fallacy mainly in two types of cases. One is where the conclusion of the slope argument is stated in such strong terms, like "inevitably," "necessarily," and so forth, that it could never be proved by an inherently presumptive, defeasible, and conjectural type of argumentation like the slippery slope.

The other is the type of case where so many of the steps required to fill in the sequence are left out, or are so poorly supported, or both, that the argument is little more than an innuendo or a scare tactic to try to exploit the timidity of the respondent. We see this very often in the *short form* type of slope argument like case 26, where we are given no idea, other than what we can fill in for ourselves, of what the intervening sequence between "work permits" and "police state" could be. Granted, it is not too difficult for us to fill in some intervening steps in this case, and for this reason, the argument does have some plausibility. But so much is left out that really it would be very difficult, or even impossible, for a rational critic to throw doubt on the intermediate steps with any precision, since the critic cannot even say, for sure, what these steps are supposed to be.

In the extreme cases, we can see that both these faults are present. In case 28, the slippery slope argument is little more than a provocation or incitement that exploits the fear of the respondents to try to shift the dialogue from negotiation to a quarrel. This slope argument is based on the unsupported assumption that the other side is wholly unreasonable and that they could never be trusted to negotiate or discuss the issue moderately.

In general, the type of analysis of the slippery slope argument given

in Walton (1992a) supports the thesis that this is very often a sophism type of fallacy involving aggressive argumentation tactics inappropriate for a context of dialogue, and in some cases, dialectical
shifts. Failure to meet the requirements of a scheme, however, can
also account for slippery slope paralogisms of various kinds.

5. The Fallacy of *Secundum Quid*

All the argumentation schemes in chapter 5 are presumptive in
nature, meaning, according to the analysis of Walton (1992c, chap. 2)
that how they function to incur commitment in a dialogue is based
on a shift in the burden of proof (see chapter 5, section 1, above).
When an assertion is put forward in a dialogue, the proponent of the
assertion becomes committed to the proposition contained in the assertion in a strong sense of incurring a burden of proof—she is committed to defending that proposition if it is challenged. With presumption, as opposed to assertion of a proposition, however, the
burden of proof shifts to the respondent. The respondent has the burden of disproving the proposition in question if he wants to reject it
as a commitment. According to the analysis of Walton (1992c, chap.
2), once both parties in a dialogue agree to accept a presumption put
forward by a proponent, then the respondent is not free to reject that
proposition as a commitment immediately or for a time unless he
can show that evidence exists to prove that the proposition is false.

Those who have now read chapters 2 and 5 will recognize that this
kind of reasoning has the negative logic characteristic of the argumentum ad ignorantiam—if you don't know that a proposition is true
(false), then you may presume that it is false (true).

At any rate, all the argumentation schemes of chapter 5 are presumptive in this sense. Presumptive reasoning may be contrasted
with deductive and inductive reasoning, both of which have a positive logic, as opposed to the negative logic of presumptive reasoning.
Presumptive reasoning is inherently *defeasible*, meaning that it is
open to exceptions that cannot be (absolutely) predicted in advance.
Presumptive reasoning is also *nonmonotonic*, meaning that the addition of new premises to an argument can change whether that argument is structurally correct (e.g., valid or invalid) or not. It is characteristic of presumptive reasoning generally that it is subject to
qualifications, so that once new information comes in relevant to
these qualified circumstances, the reasoning could be overturned or
defeated as applied to these circumstances.[5]

Secundum quid is the fallacy of neglecting qualifications. According to the analysis given in Walton (1992c, 75–80), this fallacy char-

acteristically occurs where presumptive reasoning, which is inherently defeasible and subject to exceptions, by its nature, is treated in a rigid or absolutistic way, as though it were, for example, deductive reasoning, of a kind that is monotonic and not subject to exceptions as a kind of reasoning.

Many generalizations in everyday conversation are expressed in a *generic* fashion—that is, no explicit quantifier like 'all,' 'some,' or 'many' is stated. Given a generalization like 'Ravens are black' or 'Birds fly,' it may depend on the context of dialogue whether it should be taken as a strictly universal generalization, for example, 'All ravens (without exception) are black,' or as a defeasible generalization, 'Typically, birds fly.' The strict universal generalization, which can be refuted by even a single counterexample, warrants a deductive inference, for example:

Case 107

> All ravens are black.
> Rodney is a raven.
> Therefore Rodney is black.

In contrast, the qualified or defeasible generalization warrants only a presumptive inference that is subject to default, in some instances.

Case 108

> Birds fly.
> Tweety is a bird.
> Therefore Tweety flies.

Suppose that, in an extension of the case, we find out that Tweety is a penguin, a type of bird that does not fly. The inference in case 108 is then defeated as a basis for inferring the conclusion 'Tweety flies.' Once that new information came in, that Tweety is a penguin, it is taken into account that Tweety is one of those exceptional kinds of birds to which the generalization 'Birds fly,' meaning 'Typically, birds fly,' or 'If x is a bird, we can assume by default, subject to exceptions, that x flies,' does not apply. In this case, then, the monotonic and defeasible reasoning of the inference in case 108 is subject to default. Instead we are warranted in concluding that in this case, although Tweety is a bird, he does not fly.

As we can see in case 24 and 25, the fallacy of secundum quid is the failure to recognize the nonabsolute character of defeasible generalizations and inferences that are inherently subject to default in exceptional cases. The fallacy is one of treating a defeasible, qualified generalization as though it were a strict, universal generalization. For example, the principle 'Boiling water that will be hot enough to cook an egg hard in five minutes' is defined for standard conditions one would normally be expected to encounter in a typical case where the principle would be put to use. To treat the generalization as though it must strictly apply to a nontypical case, for example, five thousand feet above sea level, would be a secundum quid fallacy if one tried to draw the conclusion absolutely that, even in this case, the water *must* be hot enough to cook an egg in five minutes.

Thus we can see that secundum quid is an error-of-reasoning type of fallacy that confuses two different types of warrants for reasoning. It is also partly a dialectical fallacy, however, because it is always a function of the context of dialogue to determine which standard of argument is appropriate, for example, whether a generalization should be interpreted as strictly universal or as defeasible.

6. The Straw Man Fallacy

The main characteristic of persuasion dialogue is that the premises of a proponent's argument must be commitments of the respondent. This is vital if the dialogue is to be truly interactive, for otherwise the participants will not be really dealing with the opinions of each other, and the conflict of opinions will not be resolved by their argumentation together.

The concept of commitment in dialogue as the basic characteristic of dialogue logic was introduced by Hamblin (1970). Hamblin saw each participant in a dialogue as having a log or tableau of propositions called a *commitment set*. As each participant makes various kinds of moves in the dialogue, propositions are inserted into or retracted from this set. In Walton (1985a) the commitment set was identified with the *position* of an arguer, representing her developing point of view, the collection of propositions she had committed herself to, during the course of a dialogue.

As noted in chapter 5, section 5, argumentation from commitment is a type of presumptive reasoning that can correctly and appropriately be used in a dialogue to shift a burden of proof. Many common types of argumentation, like the circumstantial ad hominem argument, for example, are species of argumentation from commitment.

As noted in chapter 2, section 10, the straw man fallacy occurs where one party in a dialogue misinterprets the position of the other party by exaggerating it, making it seem foolish, or otherwise distorting it so that it seems weaker and more open to refutation than it really is. As Hurley (1991, 119) puts it, "The *straw man* fallacy is committed when the arguer misinterprets an opponent's argument for the purpose of more easily attacking it, demolishes the misinterpreted argument, and then proceeds to conclude that the opponent's real argument has been demolished." This excellent description of the straw man fallacy shows that it is basically a sophistical refutation of the Aristotelian type—a sophistical tactic used by one party in a dialogue to give the appearance of having refuted the other party without the reality of it. But the fallacy is also partly an error-of-reasoning type of fault that consists in a misapplication of the argumentation scheme of the argument from commitment.

For example, the problem in case 31 is whether Jim's environmentalist position, as expressed in his dialogue with Mavis about improving the environment by controlling pollution, gives a sufficient basis of premises for drawing the conclusion that Jim is committed to making the environment "a natural paradise on earth." Confronted with this ostensible consequence of his commitment set, Jim has the option of denying his commitment to it. Mavis and Jim, then, could argue out whether Jim is really committed to this proposition or not, given the textual evidence of what he said before.

In such cases, however, we need to distinguish between an incorrect or insufficiently supported allegation by one party that another party is committed to some proposition, and the straw man fallacy, where such an allegation is used as a tactic to attack and demolish the misinterpreted position of the other party as a deceptive technique of refutation in a dialogue. A misinterpretation of a commitment can be easily corrected, but a systematic tactic of distorting someone's commitments for the purpose of refuting their argument is something else again. This can be a systematic tactic of deception used to browbeat someone and continually (and often very effectively) to make their arguments look silly, or even repulsive, in the eyes of an audience. This is a far more common tactic, especially in eristic dialogue, than is commonly recognized.

Identification and evaluation of the straw man fallacy in a given case requires a dialectical approach of examining the prior sequence of dialogue to find profiles of dialogue where the party attacked has made his commitments clear through the moves he has made in the dialogue. Of course, in real life, this is something of an idealization, because in many cases, participants do not keep track of what they said by means of a tape recorder or some other device that yields a transcript. Therefore, how one can apply the argumentation scheme of the argument from commitment in a given case depends on the information available in that case relevant to determining the arguer's commitments.

7. Equivocation

The fallacy of equivocation is not due to the failure or abuse of any single type of argumentation scheme. Indeed, an equivocal "argument" is not really one single argument at all. It is a sequence of sentences put forward in a dialogue as something that is supposed to be an argument, whereas in reality, it is a whole bundle of arguments, which only appears to be an argument because of ambiguity in a key term. Ambiguity is not, in itself, fallacious, nor is it always a wholly bad thing in dialogue. If you are supposed to be putting forward an argument in a dialogue, however, equivocation is a species of failure to fulfill this obligation. Such a failure makes it impossible for a rational critic to criticize your "argument" in the way that would be appropriate for her normally in a context of dialogue, for example, by showing that it is invalid or that the premises are not adequately supported and so forth.

In case 37, for example, the set of sentences below supposedly puts forward an argument.

> Laws of nature exist, in science.
> Whenever there is a law, there is a lawgiver—someone who creates the law.
> Therefore there exists a lawgiver who is a power above nature.

The problem is here that the first premise is only plausible if we interpret the term 'law' as the term used in science to denote scientific generalizations, expressed by equations, formulas, and the like, in scientific terms. By contrast, however, the second premise is plausible only if you interpret 'law' to mean a man-made convention, or rule of conduct adopted by a group of people. So disambiguated, however, the premises would give us no valid or structurally warranting basis for inferring the conclusion—no matter which way we interpret 'law' in the conclusion.

Once disambiguation has taken place, all we get is a set of four possible arguments, none of which individually would be of any worth as an argument (with plausible premises and an argumentation scheme, form, or structure, that would enable one to use the premises to prove or support the conclusion). Only the ambiguity makes it appear that here we have a worthy or useful argument to support a conclusion that is at issue. Equivocation, then, is a fallacy of concealment of failure to fulfill correctly a burden of proof by presenting an argument that fulfills a probative function appropriate for the context of dialogue. Once the ambiguity is revealed, this appearance collapses.

Many logic textbooks presume that ambiguity is fallacious and postulate the "fallacy of ambiguity," although they disagree on what this fallacy consists in. Black (1946, 170) takes the line that ambiguity should generally be presumed to be a defect in argumentation subject to exceptions. Wheelwright (1962, 289) postulates the *material fallacy of ambiguity*, which "occurs when two meanings of an ambiguous word or phrase are at work in an argument." Rescher (1964, 75) has two separate fallacies—the fallacy of ambiguity and the fallacy of equivocation. Fischer (1970, 265) defines the *fallacy of ambiguity* as "the use of a word or an expression which has two or more possible meanings, without sufficient specification of which meaning is intended." These accounts pose a problem of identifying equivocation as a fallacy and distinguishing it from the so-called fallacy of ambiguity. Is there one fallacy here or two?

Equivocation is not the same thing as ambiguity. Equivocation is a fallacy. Ambiguity is not a fallacy. Ambiguity can be a problem in various types of dialogue. For example, in advice-giving dialogue, if one party is trying to give instructions to another party on how to do something, an ambiguity in a key word or phrase in the instructions could be confusing or misleading, conveying the wrong instructions. But this type of failure of communication is not necessarily a fallacy.

In some contexts of dialogue, ambiguity is tolerable and even inevitable. For example, in the early stages of a critical discussion on a controversial topic of public policy or morality like the abortion issue, there is bound to be ambiguity inherent in key terms that are subject to dispute and to proposed definitions. Such ambiguity is not necessarily a bad thing, nor should it be called fallacious per se. Whether such an ambiguity is critically bad or is part of a fallacy depends initially on how it is used or exploited in the dialogue and on what type of dialogue it is.

Equivocation is a fallacy in a critical discussion because the basic function of an argument as used in a critical discussion by its proponent is to convince the respondent rationally that the conclusion should be a commitment of his because (1) the premises are commitments of his (or propositions he can be persuaded to accept) and (2) the structural link between the premises and conclusion is such that if he accepts the premises, then that gives support to his rational acceptance of the conclusion.[6] But for an argument to fulfill this function, it must be a univocal, definite *argument*, that is, it must have definite propositions as premises and conclusion and a structural link or warrant joining them. If something that purports to be an argument commits the fallacy of equivocation, it can never fulfill this function.

8. Amphiboly

Given the traditional treatment of the logic textbooks, it has been hard to take amphiboly seriously as a fallacy. Typical textbook examples are the following cases, from Engel (1982, 81) and Fischer (1970, 267), respectively.

Case 109

With her enormous nose aimed at the sky, my mother rushed toward the plane.

Case 110

Richly carved Chippendale furniture was produced by colonial craftsmen with curved legs and claw feet.

These cases seem to be grammatical errors involving misplaced modifiers that make the sentence ambiguous, suggesting an inappropriate but funny interpretation. But are such cases fallacious? For one thing, they are not arguments but only ambiguous sentences. For another thing, they are not serious errors that would fool anyone, or be worthy of warning students in informal logic courses about, as deceptive tactics that it is important to be familiar with.

After surveying the textbook standard treatment of amphiboly, Hamblin (1970, 18) laid down the following challenge: to get a good example of amphiboly as a fallacy, we would need to find "a case in which someone was misled by an ambiguous verbal construction in such a way that, taking it to state a truth in one of its senses, he came to take it to state a truth in its other sense." Hamblin concludes that none of the textbook examples he examined meets this challenge.

It would be possible to construct some sort of implicit argument structures out of cases 109 and 110, on the basis that they suggest mistaken conclusions based on Gricean implicatures of some sort. But such an attempt to find the argument implicit in this kind of case would not be very convincing unless more information on the context of dialogue were introduced. One basic failure with amphiboly then is simply a lack of adequate casework—a lack of good, well-analyzed examples that show how amphiboly is a failure of argument of a seriously deceptive kind.

Fortunately, however, the textbooks are not entirely bereft of such cases, and in fact the two examples given by Michalos (1969, 366)—cited above in chapter 2, section 13—show how amphiboly could be a serious fallacy in misleading advertising and in other contractual kinds of dialogue that often have commercial and legal implications. In fact, lawyers spend a good deal of their time in drafting and studying contracts in order to eliminate grammatical ambiguity and in arguing cases that turn on grammatical ambiguity. The fallacy of amphiboly is always a danger in such cases, but generally we need to resist the urge to label any kind of case where a problem due to grammatical ambiguity arises as an instance of this fallacy.

The following is a typical case in point.

The plaintiff's husband died as a result of a motor vehicle accident which occurred in Barbados. The bus in which the man died was transporting him, the plaintiff, and others from their hotel in Barbados to the airport at the end of their 14-day vacation. The couple had purchased the vacation package through an agent. As part of the package they purchased accident insurance under a group policy. The policy provided $45,000 in coverage for death occurring in consequence of riding in: (1) any aircraft . . . ; or (2) "any airport limousine or bus or surface vehicle substituted by the airline." The policy provided $15,000 in coverage for death arising out of the use of other public conveyances. The plaintiff argued that the words "substituted by the airline" in (2) above referred only to the words "surface vehicle."[7]

In this case, the action was allowed for fifteen thousand dollars on the grounds that the words 'substituted by the airline' referred to all the modes of transportation mentioned in clause (a).

A puzzling aspect of this case, however, is that, in contrast to cases 38 and 39, neither party was (presumably) intentionally trying to deceive anyone else or use any sophistical tactic of deception. Whoever wrote the insurance contract made the blunder of phrasing the clause in question in such a way that it appeared to be open to a legitimate interpretation other than the one that was presumably meant to be expressed. At least, it left room for argument sufficient for a legal case to be made out of it. Was this failure a fallacy or merely a blunder?

In contrast to cases 38 and 39, where it is appropriate to speak of a sophism of amphiboly being used, in case 111 it is appropriate to say that no fallacy was committed by the writer of the text of the insurance policy or any other participant in the dialogue (at least, in light of our theory of fallacy and our analysis of the fallacy of amphiboly).

True, there was a legal problem that arose and also perhaps a problem of communication. But is that a fallacy? Not necessarily. In fact, the judge supported the natural interpretation of clause (2), the one the writer of the policy intended (or so the text and context suggest). Cases like this indicate that amphiboly is a serious fallacy but that we should resist the tendency to say that all troublesome cases of grammatical ambiguity in dialogue are instances of this fallacy.

9. Accent

The worst problem with the fallacy of accent is that it has come to include a heterogeneous collection of would-be fallacies, including factors of rhetorical emphasis in the connotations of words, wrenching from context, innuendo, slanted discourse, special pleading, suppressed evidence, misquoting of sources, bias, and irony due to pronunciation emphasis. Originally, Aristotle meant accent to refer to ambiguity due to the rise and fall of intonation (in Greek). In English this is not so critical, however, as Hamblin (1970, 24) notes, leaving logicians in a dilemma of what to do with the fallacy of accent. Consequently, accent came more and more to be equated with verbal emphasis in the pronunciation of a sentence, where the concept of emphasis came to be taken more and more broadly. The result has been what Hamblin (1970, 25) described as a "slippery slide" beginning with verbal emphasis to include all kinds of emphasis and contextual balance. Starting out as a fallacy that was so narrow as to be relatively trivial, accent became so broad and heterogeneous that it no longer seems to represent any single fallacy.

The simplest kind of case of accent is the use of verbal stress in pronouncing a sentence to convey a conclusion covertly, to be drawn by the respondent in context through implicature. For example, pronouncing the sentence "We should not speak ill of our *friends*" with the indicated stress seems to suggest that it is all right to speak ill of someone who is not your friend. This could be a sophism, in that it is a way for the speaker to escape commitment if asked, "Did you really mean to say that it is all right to speak ill of some people?" The speaker can then reply, "I never said that!" It does seem possible to analyze this type of case as a sophism on the grounds that a conclusion is suggested by implicature (in the context of dialogue), but then later commitment to that proposition is (illicitly) retracted. The device is a common and effective technique used to achieve plausible deniability in argumentation.[8] Once we expand the boundaries of the fallacy of accent beyond this relatively manageable type of case, however, a lot of problems and difficult questions of classification and identification of fallacies begin to arise rapidly.

Wrenching from context is an important dialectical fallacy in its own right but is closely related to the straw man fallacy. The fallacy of wrenching from context occurs where part of a text is selected and quoted to support an argument but where a fair consideration of the wider context of the selected part quoted would not support the argument at all or not nearly as strongly as the selected part, by itself, appears to. This type of fallacy is a sophistical tactic that we are all familiar with.[9] A common type of example, cited by Damer (1980, 16), is the selective quotation of parts of book reviews, used by the publisher for promotion purposes, that makes the review of the book appear much more favorable than the whole review does. One can see that in such a case, elements of secundum quid could be involved as well if qualifications stated in the review are overlooked.

Copi and Cohen (1990, 116) and Damer (1980, 16) classify the fallacy in this type of case as that of accent. This is far too broad a way of defining the fallacy of accent, and cases of this sort should be classified under the separate heading of the fallacy of wrenching from context. Wrenching from context is a sophism that uses selective quotation of a passage out of context to support an argument in a misleading way and can be evaluated as such, once the larger context of dialogue is included.

Another major category of fallacy commonly included under the heading of accent in the textbooks is *special pleading*, defined by Robinson (1947, 191) as "emphasizing the parts of a subject matter or those arguments for or against a theory which are favorable to your own position, and omitting the parts which are unfavorable to it." This is considered a species of the fallacy of accent because the arguer is "wrongly accenting" by "stressing only part of the truth" (Hamblin 1970, 25). But is it a fallacy? Hamblin (25) notes the necessity of admitting that any partisan argumentation, for example an attorney pleading his case in court, "must be engaging in 'special pleading.' "

In fact, special pleading is really only another name for bias in argumentation. As Blair (1988) has convincingly argued, however, not all bias is bad bias. Bias should only be described as a fallacy if it is so bad in a given case that it impedes the argumentation in a dialogue from contributing to realizing the goal of the dialogue. Recent attempts to define bias in argumentation, however (Walton 1991a), have taken the line that bias is not in itself a fallacy.

However such would-be fallacies as special pleading, half-truth, suppression of evidence, and the like are dealt with, it is clear that they are better evaluated under the category of bias in argumentation. It does not help matters to include them under the heading of the fallacy of accent. Although these things are all related to the fallacy of accent in some way, it is much better to define accent in a narrower way, so that it is comparable with, and in the same general category as, the other fallacies within language of equivocation and amphiboly.

Clearly, however, a lot of work remains to be done in defining the borders of the fallacy of accent more exactly. Although bias is partly linguistic, it is not a fallacy within language—and for that matter, not even a fallacy except in certain severe cases where it may be identified with other fallacies like secundum quid.

10. Fallacies and Violations of Rules

It is tempting to think that the rules for a critical discussion can be used to classify a particular fallacy as a violation of a particular rule, so that when a specific fallacy occurs, you can say, "There, that was a violation of rule *x*, therefore it is a case of fallacy *y*." But the rules for the critical discussion given by van Eemeren and Grootendorst (1984; 1987) are very broad. They express in broad terms guidelines on the means to carry out the goals of the dialogue. For example, there is a rule that burden of proof must be fulfilled when requested. But as we have seen, most, if not all, the major fallacies involve failure to fulfill this requirement. And it is how such a failure exactly occurs, by what means, that defines which fallacy occurred, or whether a fallacy occurred. For failure to fulfill burden of proof is not itself a fallacy at all, much less any specific fallacy. What needs to be determined is what specific means were used to carry out the failure. The means are basically arguments, types of argumentation that must be used in certain ways, if the goals are to be achieved. Specific ways of misusing these arguments, ways that block the goals, are fallacies. But there is no one-to-one correspondence between the rules and these fallacies.

The ten rules for critical discussion of van Eemeren and Grootendorst (1987) do not identify individual fallacies. Instead, the major fallacies are identified with the types of argumentation characterized by their presumptive argumentation schemes. Rule 1 applies to many of the cases of fallacies, because different tactics are often used to try to prevent parties from expressing or casting doubt on a standpoint. Van Eemeren and Grootendorst themselves concede that ad hominem, ad baculum, and ad misericordiam (212–13) all violate this rule. So this rule does not single out any particular fallacy. Rules 3 and 4, in effect, stipulate that an argument must be relevant to the issue of a dialogue. Although irrelevance could be called one big fallacy of ignoratio elenchi (thus violating these two rules), many of the fallacies are in significant part, but not totally, characterizable as failures of relevance. Rule 2 expresses the idea of burden of proof. But failure to defend a thesis if we are asked to do so is not in itself a fallacy (as noted above). Nor is it identifiable with any single fallacy. Failure to back up a contention when we are asked to do so is a fault or error—it means your argument does not meet the burden-of-proof requirement and is therefore unsupported or insufficiently proved. But that in itself does not mean the argument is fallacious.

Many of the fallacies studied above, and in chapter 7, are associated, at least in part, with a failure to fulfill burden of proof. Begging the question is one; ad hominem is another. The fallacy most intimately connected with failure to fulfill burden of proof is the argumentum ad ignorantiam. But the fallacy here is the inappropriate shifting of the burden of proof onto the other party. The fallacy is not itself identical to the failure to fulfill burden of proof. It is not just a violation of rule 2, and that's how it is defined as a fallacy. It is a special type of tactic used to try to shift the burden of proof deceptively or inappropriately from one side to the other in a dialogue.

The same can be said for the fallacy of petitio principii. Although sometimes wrongly identified with the fault of failure to fulfill burden of proof, such a failure is not identical or equivalent to the fallacy of petitio. The fallacy of begging the question, or petitio principii, involves the essential use of circular argumentation, which, while not fallacious in itself, is fallaciously used to evade a proper fulfillment of burden of proof in a dialogue.

Rule 7 (and possibly with it rule 8) is a sort of granddaddy rule that covers most of the major informal fallacies (rule 8 perhaps covering the formal fallacies, depending on what is meant by 'valid'). For as we have seen, most of these fallacies are essentially arguments where the defense has not taken place by means of an appropriate argumentation scheme that has been correctly applied. This rule, then, like the others, does not equate with any single fallacy. Rather, its violation can be partially identified with many of the fallacies in some cases. It is not a characteristic, or defining condition, or an analysis, of any single fallacy.

But the rules are connected to the fallacies. The rules give you a broad insight into what went wrong with a particular fallacy with respect to its getting away from supporting the goals of a dialogue. For example, the rule "Be relevant!" can be used to explain why a particular argument that was wildly off topic by making a personal attack in the midst of a scientific inquiry is blocking the dialogue from taking its proper course. Or the rule "Fulfill the burden of proof!" may indicate what's wrong when someone keeps attacking the other party personally with wild innuendo without backing it up by any real evidence of wrongdoing. But in both cases, the fallacy might be an ad hominem fallacy. Here the rule gives you insight into what has gone wrong basically, but it does not pinpoint or identify the fallacy. It might be quite misleading to say that both these cases are instances of the ignoratio elenchi fallacy, when it would be much more specific and accurate to say that they are instances of the ad hominem fallacy.

As shown by Walton (1989a, chap. 1), there can be more general and more specific rules of dialogue, and there can be positive and negative rules. The goals link to general rules, which in turn link to more specific subrules for specific situations, and in turn, these subrules link more closely to fallacies. The following is a good example of how this system works. In a critical discussion, the goal is to resolve a conflict of opinions by giving each side the freedom and incentive to bring out its strongest arguments to support its side of the issue. For the critical discussion really to succeed, there must be a clashing of the strongest arguments on both sides and good responses by the other side when an argument is very telling against its point of view. In turn, for this to happen, there must be freedom on both sides to express one's point of view as fully as possible. Clearly, in order to function successfully, a critical discussion needs freedom to express a point of view. A certain quality of openness on both sides is required. This is a positive requirement that leads to a negative rule, namely the rule that neither side must prevent the other side from expressing its point of view.

There are all kinds of ways of violating this negative rule, however. One party may say, "If you know what's good for you, you will shut up right now!" Or one party may ask an unfair question like, "Have you stopped your usual cheating on your income tax?" Or one party may simply keep talking, out of turn, thus preventing the other party from saying anything at all. The first two tactics are two different types of fallacies, and the third is not a fallacy at all—or at least it is not specifically identifiable with any of the traditional list of fallacies. The first tactic is to make a threat that will presumably prevent the other party from putting forward any further argumentation at all. The second tactic poses a question such that, no matter which way the respondent answers it directly, he concedes a defect of veracity that prohibits him from putting forward any further arguments that will have any credibility in the discussion.

To make the first two fallacies fallacies that are so by virtue of being specific rule violations, we must invent rules like the following: "Don't make threats in a critical discussion or in any other type of dialogue where the threat is designed to prevent the other party from taking part properly in the dialogue!" or "Don't ask complex questions with presuppositions that are defeating to the respondent's side unless you get her to agree to the presuppositions using prior questions in the sequence of dialogue first!" These very specific rules are now sharp enough to characterize particular fallacies.

These do not seem to be the kinds of rules that van Eemeren and Grootendorst have in mind, however. Moreover, it is obvious that no matter how you define or characterize a particular fallacy, once you have the characterization of it, you can always make up a rule saying "Don't do that!" But this way of proceeding would be a circular way of saving the definition of fallacy as a violation of some set of rules for argumentation in dialogue. In general, then, pointing to a rule violation is not a sufficient way of either pinpointing that a particular fallacy was committed or of evaluating that argument as fallacious. We have to look elsewhere both to identify the fallacies and to define each of them as distinctive entities.

The inherent nature of fallacy, according to theory given in chapter 8, is to be found in the Gricean principle of cooperativeness, which says that you must make the kind of contribution required to move a dialogue forward at that specific stage of the dialogue. This principle requires that at any given point in a dialogue, a certain kind of sequence of moves is needed in order to make the dialogue go forward. Each participant has to take proper turns, first of all. Then once one participant has made a certain type of move, like asking a question, the other party must make a move that matches the previous move, like providing an appropriate response. A set of these matching moves and countermoves is a connected sequence that makes up a profile of dialogue. This profile, when viewed in the context of dialogue, identifies the fallacy that occurred.

Fallacies come into a dialogue essentially because the profile gets balled up in a way that is obstructive. The one party tries to move ahead too fast by making an important move that is not yet proper in the sequence. Or the one party tries to shut the other party up by closing off the dialogue prematurely or by shifting to a different type of dialogue. In such cases, the sequence may start out right, but then the moves start to happen in the wrong places in the sequence. Or key moves are left out of a sequence that should have been properly put in. The result is that the sequence is not in the right order required for that type of dialogue and at that particular stage of the dialogue. This is where a fallacy occurs, where the resulting disorder is a type of sequence that blocks the dialogue or impedes it seriously.

For example, in the case of a fallacious argumentum ad ignorantiam, the one party may put forth an assertion he has not proved, or has even given any argumentation for at all, and may then demand that the other party either accept it or disprove it. Here each individual move in the dialogue is all right, but what has gone wrong is that the first party failed to give some support to his argument before making his move of demanding that the other party accept or disprove it. The fault here was the key missing move in the sequence.

Of course, you could say that this case was simply a failure to fulfill burden of proof, which is, of course, a violation of a rule of a critical discussion. But that, in itself, was not the fallacy. Mere failure to prove something is not itself a specific fallacy per se. What went wrong was the failure to do what was required at the right step in the sequence. Such a fallacy is adequately modeled normatively only by looking at the whole sequence of moves and seeing that one required move was missing. The profile of dialogue reveals the fallacy, not the single missing move by itself.

Another case is the fallacy of begging the question. Again, the failure is one of an arguer trying to push ahead too aggressively in a dialogue by balling up the proper sequence. Instead of fulfilling burden of proof properly by an appropriate sequence of argumentation, the proponent tries to conceal this failure by pressing in a proposition that is in doubt as a premise. Once again, the fallacy is not simply the violation of fulfilling the requirement of burden of proof, although that was part of it. The fallacy can be identified only by looking at the whole sequence of argumentation, which could be done by using an argument diagram, or a profile of dialogue, and ascertaining that the sequence in the profile is of a circular configuration. That is, it comes back to the same point or proposition previously in the sequence already. Then this profile has to be shown to be inappropriate for the given stage and context of dialogue. The actual profile has to be shown to fall short of the correct type of profile for that stage of a normative model of dialogue.

Of course, it is informative to say that such a sequence is wrong because a rule of a specific type of dialogue, like a critical discussion, has been broken. But that in itself is not sufficient to explain why a fallacy occurred or to determine which of the fallacies was committed. To do that, one has to look at the profile of dialogue and see how the tactic that was used balled up the right sequence of dialogue in a particular way, in order to identify the sophism.

With respect to some of the twenty-five major informal fallacies, in particular, where the fallacy is a paralogism, the order in the profile is determined by the kind of argumentation scheme that is appropriate, the accompanying critical questions matching that scheme. In these cases, identification of which fallacy has been committed can be carried out by identifying the argumentation scheme that was used. But this procedure works for only some cases of fallacies.

For example, if the argumentation scheme that was used was the negative argument from ethos, then if a fallacy occurred through the wrong use of this scheme in context, we can say that the fallacy that occurred was the abusive ad hominem.

But identifying or classifying a fallacy is different from evaluating or explaining a fallacy. If too little or no evidence was given to support the premise that a person has a bad character, would an ad hominem argument then be fallacious? Maybe not, if when asked to supply such evidence, the arguer complied or at least did not try to evade the request or show other evidence in the profile of dialogue of making inappropriate further moves.

Evaluating whether a particular case is fallacious or not, especially where the fallacy is a sophism, requires essential reference to the wider profile of dialogue. Knowing that the argumentation scheme was used incorrectly is not, in itself, sufficient for such a determination. The reason is that if an argumentation scheme was used incorrectly, it could have been a slip or oversight. Much may depend on the kind of follow-up moves made in response to the other party's critical questioning of the move. It does not follow, in every instance, that a fallacy was committed. The reason is that there is a difference between an error in argumentation and a fallacy. A fallacy is a particularly serious kind of error, or an infraction of the rules of dialogue, identified with a baptizable type of argumentation that has been abused in such a way as to impede the goals of the type of dialogue the participants in the argumentation were rightly supposed to be engaged in.

Notes

1: The Concept of Fallacy

1. The term 'baptizable' is due to Johnson (1987). A baptizable error is one that is common enough and serious enough to merit naming as a fallacy.

2. In their paper "The Current State of Informal Logic," J. Anthony Blair and Ralph H. Johnson (1988) identified "the theory of fallacy" as an important lacuna in the development of informal logic.

3. The history of the subject of fallacies has been presented very well by Hamblin (1970).

4. Walton (1987) and van Eemeren and Grootendorst (1987).

5. Hamblin (1970, chap. 1).

6. Hamblin (1970, chap. 8). See also Hamblin (1971).

7. See Blair (1993) for an interesting discussion of some potential conflicts in these rules.

8. Willard (1989) has made the same general point that the types of argument identified with the fallacies are sometimes reasonable. See also the remarks of Johnson (1993).

9. Van Eemeren and Grootendorst (1987, 287).

10. This approach has already been advocated in Walton (1989a).

11. Van Eemeren and Grootendorst also advocate the use of argumentation schemes but provide no systematic account of them.

12. See Walton (1989b, 67–71) for a more extensive analysis of this profile of dialogue.

13. See section 7 above, on persuasion dialogue.

2: Informal Fallacies

1. This case is extensively analyzed in Walton (1985a).
2. On defining 'bias,' see Blair (1988) and Walton (1991b).
3. Copi (1982, 99) defines it as appeal to force or the threat of force. Engel (1976, 130) calls it the "fallacy of appeal to fear."
4. Walton (1989a, chap. 7) gives a range of typical cases.
5. This fallacy is treated extensively in Walton (1989a, chap. 2) where many examples are given. Of special interest is the common nature of these types of questioning tactics in parliamentary debates.
6. The problem here could be partly linguistic. See the account of the use of loaded terms in section 13 of this chapter.
7. See Walton (1989a, chap. 7).
8. This idea of evidential priority can be found in Aristotle. See Hamblin (1970) and Walton (1991a).
9. See also Colwell (1989) for an extensive discussion of this type of case.
10. This type of case has been much discussed by statisticians. See Walton (1989a, chap. 7).
11. Walton (1989a, 229).
12. For ease of exposition, this definition is given only for deductively valid arguments. But it can be straightforwardly extended to inductive and presumptive arguments, using the concept of an argumentation scheme.
13. Manicas and Kruiger (1968, 331) and Byerly (1973, 45), for example.
14. This is a moot point, however. Vagueness and ambiguity have often been classified as fallacies, and the issue does challenge what is meant by 'fallacy' generally. Often it has been presumed or required that a fallacy should be some sort of fallacious argument, in textbook treatments.

3: Formal Fallacies

1. Aldrich (1862).
2. See Hamblin (1970, chap. 6).
3. Examples are given in this chapter, below.
4. See chapter 2 on the fallacy of hasty generalization.
5. See also Hughes and Cresswell (1968, 27) on scope confusion ambiguities in applying modal logic to natural language.
6. See chapter 2 on linguistic fallacies.
7. But see Krabbe (1990) for a different point of view. According to Krabbe, it depends on the situation whether it should be correct to frown on inconsistency as a weakness or blunder in argumentation, but we should not go so far as to count inconsistency among the fallacies.
8. Walton (1985a, 66–68).
9. See also Barth and Martens (1977) and Walton (1985a).

10. This passage from Locke is quoted in full in Hamblin (1970, 159–60).

11. See chapter 2 on the ad hominem fallacy.

4: Types of Dialogue

1. These types of dialogue are also described in Walton (1989b, 354–60), Walton (1990a), Walton (1991a, 36–45), Walton (1992c), and Walton and Krabbe (1995).

2. See note 1.

3. This idea was characteristic of Hamblin's general approach and was reflected in the conception of persuasion dialogue presented in Walton (1984).

4. Walton and Krabbe (1995).

5. Walton (1989a).

6. Walton (1984, 247–55).

7. See Walton (1992a).

8. Walton and McKersie (1965, 5).

9. This fact has important implications for the analysis of the ad baculum fallacy.

10. Woods and Walton (1989, 154).

11. Walton (1990, 416) draws a parallel between the inquiry and the Aristotelian demonstration.

12. Woods and Walton (1989, 153–58).

13. Reprinted in Woods and Walton (1989, chap. 10).

14. See Walton (1991a).

15. Thus some communication theorists think of the quarrel not as a normative model of dialogue but merely as a type of discourse event. In contrast, some authors have described the quarrel as a normative model—see Kotarbinski (1963) and Flowers, McGuire, and Birnbaum (1982).

16. The quarrel, as the saying goes, generates more heat than light.

17. On the various functions of "why" questions in dialogue, see Hamblin (1970, 273–74).

18. Walton (1990a).

19. Ibid.

20. Quoted in *Newsweek*, September 28, 1992, 61.

5: Argumentation Schemes

1. Walton (1991a).

2. Ibid.

3. Inductive reason could also possibly be described as "provisional" in nature. Many would characterize plausible or presumptive reasoning as a species of inductive reasoning. Not wishing to exclude this possibility, we would still maintain that there is a fundamental distinction to be made, at the pragmatic level, between inductive and presumptive reasoning.

4. See the verbal slippery slope argument in section 10 below.

5. Ibid.

6: Dialectical Relevance of Argumentation

1. A recent issue of the journal *Argumentation* (vol. 6, no. 2, May 1992) was devoted to relevance. The consensus noted in the editor's introduction, 138, is that the articles in the issue reflect "a growing tendency towards agreement with respect to the definition of relevance as a context-dependent pragmatic notion."

2. See Walton (1987).

3. Ibid.

4. Hamblin (1970, 256) pointed out that dialectical systems can be studied both formally and descriptively.

5. See van Eemeren (1986).

6. See Edmondson (1981, 38) on turn taking in conversational dialogue.

7. Hamblin (1971).

8. Moore, Levin, and Mann (1977).

9. Walton (1985a).

10. See Hamblin (1971) and Manor (1981).

11. Moore (1986).

12. See Hamblin (1971) and Walton (1987).

13. See the analysis of presumption in chapter 2, section 5.

14. Walton (1982).

15. See Walton (1982) as well.

16. Epstein (1990).

17. Ibid.

18. See Walton (1987).

19. Some cases of this type are studied in sections 5 and 6 below, and other cases are studied in Edmondson (1981), Manor (1982), Walton (1982), Sanders (1987) and Walton (1987).

20. Manor (1982, 72). See also the comments in section 5 below.

21. See Manor (1981).

22. See Walton (1987).

23. Ilbert (1960).

24. On probative relevance in law, see Ilbert (1960).

25. Other cases are given in Walton (1982).

26. Walton (1985a).

27. Weber (1981, 308).

28. In Walton (1992d) it is argued that four of the major informal fallacies—ad baculum, ad hominem, ad misericordiam, and ad populum—are fallacies to an especially prominent extent because they are failures of dialectical relevance in argumentation. Other fallacies are more peripherally related to failures of relevance, and still others are only tangentially related to failures of relevance.

7: A New Approach to Fallacies

1. See chapter 2, section 7.
2. See the accounts of these fallacies in chapter 2.
3. Profiles for this fallacy and other question-asking fallacies have been constructed in Walton (1989b).
4. See the comparable analysis in Walton (1989b).
5. Compare Walton (1991a, 3–4, 138, and 290).
6. See Walton (1991a, chap. 5).
7. See chapter 2, section 4.
8. See chapter 9, section 2.
9. See chapter 7, section 8.
10. See chapter 2, section 2, and chapter 7, section 9.
11. See chapter 2, section 7, and chapter 9, section 5.
12. See chapter 9, sections 7 and 8.
13. See chapter 2, section 10, and chapter 9, section 6.
14. Donohue (1981a).
15. See Krabbe and Walton (1993).
16. In chapter 8, this requirement is weakened somewhat. To be a fallacy, more carefully speaking, something must be an argument, or at least it must have been brought forward in a context and situation dialogue where an argument was supposed to be presented.

8: A Theory of Fallacy

1. See chapter 1, section 1.
2. Ackermann (1970, 47).
3. Ibid., p. 308.
4. Ibid., p. 531.
5. *De sophisticis elenchis*, p. 11.
6. Ehninger (1970, 104).
7. Walton and McKersie (1965).

9: Putting the Theory to Work

1. See also Walton (1992b, chap. 5).
2. See also the analysis of argumentation from consequences in Walton (1992a).
3. Ibid.
4. Ibid.
5. See also the account in Walton (1992c).
6. See Walton and Krabbe (1995).
7. Gorgichuk v. American Home Assurance Co., CCH DRS 1985 P43-004, O. 1985, I.L.R. P1-1984, Ontario (S.C.), April 19, 1985.

8. Even here, however, each case needs to be examined on its merits, and the context of dialogue is generally very important. The problem here is one of fixing commitment in dialogue, in the sense of pinning down an arguer to a specific commitment she has asserted (usually indirectly, by presumption).

9. It often occurs in connection with uses of sources in appeal to expert opinion in argumentation.

References

Ackermann, Alfred S. E. 1970. *Popular Fallacies*. 4th ed. Detroit: Gale Research.

Aldrich, Henry. 1862. *Artis logicae compendum*. [1691.] Edited by H. L. Mansel as *Artis logicae rudimenta* [1849]. 4th ed. Oxford: Hammans.

Aristotle. 1928. *The Works of Aristotle Translated into English*. Edited by W. D. Ross. Oxford: Oxford University Press.

———. 1955. *On Sophistical Refutations*. [*De sophisticis elenchis*]. Translated by E. S. Forster. Loeb Classical Library Edition. Cambridge, Mass.: Harvard University Press.

———. 1968. *The Nicomachean Ethics*. Translated by H. Rackham. Loeb Classical Library. Cambridge, Mass.: Harvard University Press.

Arnauld, Antoine. 1964. *La logique; ou, L'art de penser*. [1662]. Edited by James Dickoff and Patricia James. *The Art of Thinking*. New York: Bobbs-Merrill.

Audi, Robert. 1989. *Practical Reasoning*. New York: Routledge.

Bailey, F. G. 1983. *The Tactics of Passion*. Ithaca: Cornell University Press.

Bar-Hillel, Yehoshua. 1964. "More on the Fallacy of Composition." *Mind* 73:125–26.

Barth, E. M., and E. C. W. Krabbe. 1982. *From Axiom to Dialogue*. New York: De Gruyter.

Barth, E. M., and J. L. Martens. 1977. "*Argumentum ad Hominem*: From Chaos to Formal Dialectic." *Logique et Analyse* 77–78:76–96.

Beauchesne's Parliamentary Rules and Forms. 1978. 5th ed. Edited by Alistair Fraser, G. A. Birch, and W. F. Dawson. Toronto: Carswell.

Bentham, Jeremy. 1969. "The Book of Fallacies." [1984]. In *A Bentham Reader*. Edited by Mary P. Mack, 331–58. New York: Pegasus.

Black, Max. 1946. *Critical Thinking*. New York: Prentice-Hall.

Blair, J. Anthony. 1988. "What is Bias?" In *Selected Issues in Logic and Communication*. Edited by Trudy Govier, 93–103. Belmont, Calif.: Wadsworth.

———. 1993. "Dissent in Fallacyland, Part 1: Problems with van Eemeren and Grootendorst." In *Argument and the Postmodern Challenge*. Edited by Raymie E. McKerrow, 188–90. Annandale, Va.: Speech Communication Association.

Blair, J. Anthony, and Ralph H. Johnson. 1988. "The Current State of Informal Logic." Paper Presented at the Twenty-eighth World Congress of Philosophy, Brighton, England, August 21.

Brinton, Alan. 1985. "A Rhetorical View of the *ad Hominem*." *Australasian Journal of Philosophy* 63:50–63.

———. 1986. "Ethotic Argument." *History of Philosophy Quarterly* 3:245–58.

Broad, William, and Nicholas Wade. 1982. *Betrayers of the Truth*. New York: Simon & Schuster.

Brockhaus Enzyklopädie. 1974. Wiesbaden: F. A. Brockhaus.

Burge, Tyler. 1977. "A Theory of Aggregates." *Noûs* 11:97–118.

Byerly, Henry C. 1973. *A Primer of Logic*. New York: Harper & Row.

Campbell, Stephen K. 1974. *Flaws and Fallacies in Statistical Thinking*. Englewood Cliffs, N.J.: Prentice-Hall.

Canada. House of Commons. 1979. *Debates*. November 30.

———. 1984. *Debates*. February 16.

Capaldi, Nicholas. 1971. *The Art of Deception*. Buffalo: Prometheus Books.

Carney, James D., and Richard K. Scheer. 1964. *Fundamentals of Logic*. New York: Macmillan.

Carroll, Robert. 1983. "Fallacies and Argument Analysis." *Informal Logic* 5:22–23.

Castell, Alburey. 1935. *A College Logic*. New York: Macmillan.

Cederblom, Jerry, and David W. Paulsen. 1982. *Critical Reasoning: Understanding and Criticizing Arguments and Theories*. Belmont, Calif.: Wadsworth.

Chazin, Susan. 1989. "Learning to Appreciate the Dandelions in Life." *New York Times*, June 25, p. 32.

Clarke, D. S., Jr. 1985. *Practical Inferences*. London: Routledge & Kegan Paul.

Cohen, L. Jonathan. 1977. *The Probable and the Provable*. Oxford: Clarendon Press.

———. 1982. "Are People Programmed to Commit Fallacies?" *Journal for the Theory of Social Behaviour* 12:251–74.

Collins, Allan, Eleanor H. Warnock, Nelleke Aiello, and Mark L. Miller. 1975. "Reasoning from Incomplete Knowledge." In *Representation and Understanding: Studies in Cognitive Science*. Edited by Daniel G. Bobrow and Allan Collins, 383–415. New York: Academic Press.

Colwell, Gary. 1989. "God, the Bible, and Circularity." *Informal Logic* 11:61–73.

Copi, Irving M. 1982. *Introduction to Logic*. 6th ed. New York: Macmillan.

———. 1986. *Introduction to Logic*. 7th ed. New York: Macmillan.

Copi, Irving M., and Carl Cohen. 1990. *Introduction to Logic.* 8th ed. New York: Macmillan.

Damer, T. Edward. 1980. *Attacking Faulty Reasoning.* Belmont, Calif.: Wadsworth. [2d ed., 1987]

Dascal, Marcelo. 1977. "Conversational Relevance." *Journal of Pragmatics* 1:309–28.

de Cornulier, Benoit. 1988. "Knowing Whether, Knowing Who, and Epistemic Closure." In *Questions and Questioning.* Edited by Michel Meyer, 182–92. Berlin: Walter de Gruyter.

de Kruif, Paul. 1932. *Men Against Death.* New York: Harcourt, Brace.

DeMorgan, Augustus. 1847. *Formal Logic.* London: Taylor & Walton.

Donohue, William A. 1981a. "Development of a Model of Rule Use in Negotiation Interaction." *Communication Monographs* 48:106–20.

———. 1981b. "Analyzing Negotiation Tactics: Development of a Negotiation Interact System." *Human Communication Research* 7:237–87.

Duden: Das grosse Wörterbuch der Deutschen Sprache. 1981. Mannheim: Bibliographisches Institut.

Edmondson, Willis. 1981. *Spoken Discourse: A Model for Analysis.* New York: Longmans.

Edwards, Paul. 1967. "Common Consent Arguments for the Existence of God." *Encyclopedia of Philosophy.* Vol. 2. Edited by Paul Edwards, 147–55. New York: Macmillan.

Ehninger, Douglas. 1970. "Argument as Method: Its Nature, Its Limitations, and Its Uses." *Speech Monographs* 37 (June), 104.

Engel, S. Morris. 1976. *With Good Reason: An Introduction to Informal Fallacies.* New York: St. Martin's Press. [2d ed., 1982]

Epstein, Richard L. 1990. *The Semantic Foundations of Logic.* Vol. 1. *Propositional Logics.* Dordrecht: Kluwer.

Evans, J. D. G. 1977. *Aristotle's Concept of Dialectic.* London: Cambridge University Press.

Fearnside, W. Ward, and William B. Holther. 1959. *Fallacy: The Counterfeit of Argument.* Englewood Cliffs, N.J.: Prentice-Hall.

Fischer, David Hackett. 1970. *Historians' Fallacies.* New York: Harper & Row.

Flowers, Margot, Rod McGuire, and Lawrence Birnbaum. 1982. "Adversary Arguments and the Logic of Personal Attacks." In *Strategies for Natural Language Processing.* Edited by Wendy G. Lehnert and Martin H. Ringle, 275–94. Hillsdale, N.J.: Lawrence Erlbaum Associates.

Fogelin, Robert J. 1987. *Understanding Arguments.* New York: Harcourt Brace.

Franks, C. E. S. 1985. "The 'Problem' of Debate and Question Period." In *The Canadian House of Commons.* Edited by John C. Courtney, 1–19. Calgary: University of Calgary Press.

Freeman, James B. 1988. *Thinking Logically.* Englewood Cliffs, N.J.: Prentice-Hall.

Frye, Albert M., and Albert W. Levi. 1969. *Rational Belief.* New York: Greenwood Press.

Gay, Katherine. 1992. "Interview Questions That Dig Deeper." *Financial Post*, February 24, p. 522.

Glare, P. G. W. (ed.). 1982. *Oxford Latin Dictionary*. Oxford: Clarendon Press.

Golding, Martin P. 1984. *Legal Reasoning*. New York: Knopf.

Gorgichuk v. American Home Assurance Co., CCH DRS 1985 P43–004, O. 1985, I. L. R. Pl–1984, Ontario (S. C.), April 19, 1985.

Govier, Trudy. 1982. "What's Wrong with Slippery Slope Arguments?" *Canadian Journal of Philosophy* 12:303–16.

———. 1987. *Problems in Argument Analysis and Evaluation*. Dordrecht: Foris Publications.

Grice, H. Paul. 1975. "Logic and Conversation." In *The Logic of Grammar*. Edited by Donald Davidson and Gilbert Harman, 64–75. Encino, Calif.: Dickenson.

Hamblin, Charles L. 1970. *Fallacies*. London: Methuen. [Reprinted by Vale Press, Newport News, Virginia, 1986]

——— 1971. "Mathematical Models of Dialogue." *Theoria* 37:130–55.

Hardin, Garrett. 1985. *Filters Against Folly*. New York: Viking Penguin.

Hastings, Arthur. 1963. *A Reformulation of the Modes of Reasoning in Argumentation*. Ph.D. Diss. Evanston, Ill.: Northwestern University.

Hintikka, Jaakko. 1981. "The Logic of Information-Seeking Dialogues: A Model." In *Konzepte der Dialektik*. Edited by Werner Becker and Wilhelm K. Essler, 212–31. Frankfurt: Vittorio Klostermann.

———. 1987. "The Fallacy of Fallacies." *Argumentation* 1:211–38.

Hughes, G. E., and M. J. Cresswell. 1968. *An Introduction to Modal Logic*. London: Methuen.

Hurley, Patrick J. 1991. *Logic*. 4th ed. Belmont, Calif.: Wadsworth.

Ilbert, Sir Courtenay. 1960. "Evidence." *Encyclopaedia Britannica*. 11th ed., vol. 10:11–21.

Infante, Dominic A., and Charles J. Wigley. 1986. "Verbal Aggressiveness: An Interpersonal Model and Measure." *Communication Monographs* 53:61–69.

Johnson, Ralph H. 1987. "The Blaze of Her Splendors: Suggestions about Revitalizing Fallacy Theory." *Argumentation* 1:239–54.

———. 1993. "Dissent in Fallacyland, Part 2: Problems with Willard." In *Argument and the Postmodern Challenge*. Edited by Raymie E. McKerrow, 191–93. Annandale, Va.: Speech Communication Assocation.

Johnson, Ralph H., and J. Anthony Blair. 1977. *Logical Self-Defense*. Toronto: McGraw-Hill Ryerson. [2d ed., 1983]

Jones, Trevor (ed.). 1967. *Harrap's Standard German and English Dictionary*. London: George G. Harrap.

Joseph, H. W. B. 1916. *An Introduction to Logic*. 2d ed. Oxford: Clarendon Press.

Kapp, Ernst. 1942. *Greek Foundations of Traditional Logic*. New York: Columbia University Press.

Kienpointner, Manfred. 1992. *Alltagslogik: Struktur und Funktion von Argumentationmustern*. Stuttgart: Fromman-Holzboog.

Kilgore, William J. 1968. *An Introductory Logic*. New York: Holt, Rinehart & Winston.

Klein, Ernest. 1971. *A Comprehensive Etymological Dictionary of the English Language.* Amsterdam: Elsevier Scientific.

Koppel, Ted. 1988. [Interview with Michael Dukakis]. *Newsweek*, November 7, p. 53.

Kotarbinski, Thadée. 1963. "L'éristique: Cas particulier de la théorie de la lutte." *Logique et Analyse* 6:19–29.

Krabbe, Erik C. W. 1985. "Formal Systems of Dialogue Rules." *Synthese* 63:295–328.

———. 1990. "Inconsistent Commitments and Commitment to Inconsistencies." *Informal Logic* 12:33–42.

———. 1992. "So What? Profiles of Relevance Criticism in Persuasion Dialogue." *Argumentation* 6:271–83.

Krabbe, Erik C. W., and Douglas Walton. 1993. "It's All Very Well for You to Talk: Situationally Disqualifying *ad Hominem* Attacks." *Informal Logic.* 15:79–91.

Kripke, Saul. 1965. "Semantical Analysis of Intuitionistic Logic I." In *Formal Systems and Recursive Functions*, ed. J. N. Crossley and M. Dummett. Amsterdam: North-Holland.

Lambert, Karel, and William Ulrich. 1980. *The Nature of Argument.* New York: Macmillan.

Leeman, Richard. 1991. *The Rhetoric of Terrorism and Counterterrorism.* New York: Greenwood Press.

Lewis, Charlton T., and Charles Short. 1969. *A Latin Dictionary.* Oxford: Clarendon Press.

Locke, John. 1961. *An Essay Concerning Human Understanding.* Edited by John W. Yolton. 2 vols. London: Dent. [Originally published in 1690]

Lorenzen, Paul. 1969. *Normative Logic and Ethics.* Mannheim: Bibliographisches Institut.

McAuliffe, Gerald. 1980. "Just How Ethical Are the News Media?" *Quest*, February–March, 51–58.

Mackenzie, J. D. 1980. "Why Do We Number Theorems?" *Australasian Journal of Philosophy* 58:135–49.

———. 1981. "The Dialectics of Logic." *Logique et Analyse* 94:159–77.

———. 1990. "Four Dialogue Systems." *Studia Logica* 44:567–83.

Mackie, J. L. 1967. "Fallacies." *The Encyclopedia of Philosophy.* Vol. 3. Edited by Paul Edwards, 169–79. New York: Collier Macmillan.

Manicas, Peter T., and Arthur N. Kruger. 1968. *Essentials of Logic.* New York: American Book.

Manor, Ruth. 1981. "Dialogues and the Logics of Questions and Answers." *Linguistische Berichte* 73:1–28.

———. 1982. "Pragmatics and the Logic of Questions and Assertions." *Philosophica* 29:45–96.

Massey, Gerald. 1975. "Are There Any Good Arguments That Bad Arguments Are Bad?" *Philosophy in Context* 4:61–77.

Michalos, Alex C. 1969. *Principles of Logic.* Englewood Cliffs, N.J.: Prentice-Hall.

———. 1970. *Improving Your Reasoning.* Englewood Cliffs, N.J.: Prentice-Hall.

Moore, Christopher W. 1986. *The Mediation Process*. San Francisco: Jossey-Bass.

Moore, James A., James E. Levin, and William C. Mann. 1977. "A Goal-Oriented Model of Human Dialogue." *American Journal of Computational Linguistics* 67:1–54.

Moore, W. Edgar, Hugh McCann, and Janet McCann. 1985. *Creative and Critical Thinking*. 2d ed. Boston: Houghton Mifflin.

The Oxford English Dictionary. 1970. Vol. 4. Oxford: Clarendon Press.

Perelman, Chaim, and L. Olbrechts-Tyteca. 1969. *The New Rhetoric: A Treatise on Argumentation*. Translated by J. Wilkinson and P. Weaver. 2d ed. Notre Dame: University of Notre Dame Press. [First published in 1958 as *La nouvelle rhetorique: Traite de l'argumentation*]

Press, Aric, Terry E. Johnson, and Ray Anello. 1987. "A Trial That Wouldn't End." *Newsweek*, June 29, pp. 20–21.

Rescher, Nicholas. 1964. *Introduction to Logic*. New York: St. Martin's Press.

———. 1987. "How Serious a Fallacy Is Inconsistency?" *Argumentation* 1:303–16.

Robinson, Daniel S. 1947. *The Principles of Reasoning*. New York: D. Appleton-Century.

Rovere, Richard H. 1959. *Senator Joe McCarthy*. New York: Harcourt, Brace.

Rowe, William L. 1962. "The Fallacy of Composition." *Mind* 71:87–92.

Runes, Dagobert D. (ed.). 1964. *Dictionary of Philosophy*. Paterson, N.J.: Littlefield, Adams.

Salmon, Merrilee H. 1984. *Introduction to Logic and Critical Thinking*. San Diego: Harcourt Brace Jovanovich.

Salmon, Wesley. 1963. *Logic*. Englewood Cliffs, N.J.: Prentice-Hall. [3d ed., 1984]

Samuelson, Robert J. 1992. "The Dilemma of Democracy." *Newsweek*, April 13, p. 51.

Sanders, Robert E. 1987. *Cognitive Foundations of Calculated Speech*. Albany: State University of New York Press.

Sanford, David. 1989. *If P, Then Q: Conditionals and the Foundations of Reasoning*. London: Routledge.

Schopenhauer, Arthur. 1951. "The Art of Controversy." [1851]. *Essays from the Parerga and Paralipomena*. Translated by T. Bailey Saunders, 5–38. London: Allen & Unwin.

Schwarzkopf, H. Norman. 1992. *It Doesn't Take a Hero*. New York: Bantam/Linda Grey.

Seligmann, Jean, Emily Yoffe, and Mary Hager. 1991. "The Hazards of Silicone." *Newsweek*, April 29, p. 56.

Stebbing, L. Susan. 1939. *Thinking to Some Purpose*. Harmondsworth: Penguin.

Thomas, Stephen N. 1970. "A Modal Muddle." In *Free Will and Moral Responsibility*. Edited by Gerald Dworkin, 141–48. Englewood Cliffs, N.J.: Prentice-Hall.

Thouless, Robert H. 1930. *Straight and Crooked Thinking*. London: English Universities Press.

Tversky, A., and D. Kahneman. 1971. "Belief in the Law of Small Numbers." *Psychological Bulletin* 76:105–10.

———. 1982. "Judgment Under Uncertainty: Heuristics and Biases." In *Judgment Under Uncertainty: Heuristics and Biases*. Edited by D. Kahneman, P. Slovic, and A. Tversky. Cambridge: Cambridge University Press.

20/20. "It's Really a Commercial." ABC News Transcript. New York. Journal Graphics. Show #1039:12–17.

U.S. Congress. *Annals*. Washington, D.C.: Gales & Seaton, 1853.

van Eemeren, Frans H. 1986. "Dialectical Analysis as a Normative Reconstruction of Argumentative Discourse." *Text* 6:1–16.

van Eemeren, Frans H., and Rob Grootendorst. 1984. *Speech Acts in Argumentative Discussions*. Dordrecht: Foris Publications.

———. 1987. "Fallacies in Pragma-Dialectical Perspective." *Argumentation* 1:283–301.

———. 1989. "A Transition Stage in the Theory of Fallacies." *Journal of Pragmatics* 13:99–109.

———. 1992. *Argumentation, Communication, and Fallacies*. Hillsdale, N.J.: Lawrence Erlbaum.

Walton, Douglas N. 1982. *Topical Relevance in Argumentation*. Amsterdam: John Benjamins.

———. 1984. *Logical Dialogue-Games and Fallacies*. Lanham, Md.: University Press of America.

———. 1985a. *Arguer's Position*. Westport, Conn.: Greenwood Press.

———. 1985b. "Are Circular Arguments Necessarily Vicious?" *American Philosophical Quarterly* 22:263–74.

———. 1987. *Informal Fallacies*. Amsterdam: John Benjamins.

———. 1989a. *Informal Logic*. Cambridge: Cambridge University Press.

———. 1989b. *Question-Reply Argumentation*. New York: Greenwood Press.

———. 1990a. *Practical Reasoning*. Savage, Md.: Rowman & Littlefield.

———. 1990b. "Ignoring Qualifications (*Secundum Quid*) as a Subfallacy of Hasty Generalization." *Logique et Analyse* 129–30:113–54.

———. 1991a. *Begging the Question: Circular Reasoning as a Tactic of Argumentation*. New York: Greenwood Press.

———. 1991b. "Bias, Critical Doubt, and Fallacies." *Argumentation and Advocacy* 28:1–22.

———. 1992a. *Slippery Slope Arguments*. Oxford: Oxford University Press.

———. 1992b. *The Place of Emotion in Argument*. University Park, Pa.: Pennsylvania State University Press.

———. 1992c. *Plausible Argument in Everyday Conversation*. Albany: State University of New York Press.

———. 1992d. "Nonfallacious Arguments from Ignorance." *American Philosophical Quarterly* 29:381–87.

———. 1994a. *Arguments from Ignorance*. University Park, Pa.: Pennsylvania State University Press.

———. 1994b. "Appeal to Pity: A Case Study of the *Argumentum ad Misericordiam*." *Argumentation*. Forthcoming.

Walton, Douglas N., and Erik C. W. Krabbe. 1995. *Commitment in Dialogue.* Albany: State University of New York Press.

Walton, R. E., and R. B. McKersie. 1965. *A Behavioral Theory of Labor Negotiations.* New York: McGraw-Hill.

Weber, O. J. 1981. "Attacking the Expert Witness." *Federation of Insurance Counsel Quarterly* 31:299–313.

Whately, R. 1836. *Elements of Logic.* New York: William Jackson.

Wheeler, Michael. 1976. *Lies, Damn Lies, and Statistics: The Manipulation of Public Opinion in America.* New York: Dell.

Wheelwright, Philip. 1962. *Valid Thinking: An Introduction to Logic.* New York: Odyssey Press.

Willard, Charles. 1989. *A Theory of Argumentation.* Tuscaloosa: The University of Alabama Press.

Wilson, Thomas. 1552. *The Rule of Reason.* London: Grafton.

Woods, John, and Douglas Walton. 1977. "Composition and Division." *Studia Logica* 36:381–406. [Reprinted in John Woods and Douglas Walton. 1989. *Fallacies: Selected Papers, 1972–1982,* pp. 93–119. Dordrecht: Foris Publications]

———. 1978. "The Fallacy of *Ad Ignorantiam.*" *Dialectica* 32:87–99.

———. 1982. *Argument: The Logic of the Fallacies.* Toronto: McGraw-Hill Ryerson.

———. 1989. *Fallacies: Selected Papers, 1972–1982.* Dordrecht: Foris Publications.

Wreen, Michael J. 1988. "May the Force Be with You." *Argumentation* 2:425–40.

Index

Abortion, 38–39, 59, 139–40, 213, 224, 285, 293

Ad baculum, 11, 59, 211, 254, 269; defined, 40

Ad hominem, 110, 149, 162, 169, 191, 197, 267, 299; abusive (direct), 36–38, 212, 216, 229, 301; argument, 16–17, 26, 28, 30, 83, 127; attack, 166, 178, 189; basic types of, 212; bias, 37, 39, 212–13, 218, 229; circumstantial, 37, 39, 179, 212, 218, 229, 265; fallacy, 11–13, 83, 215; tu quoque, 212–14, 218; two basic problems, 39

Ad ignorantiam, 25, 162, 197, 267, 273, 301; critical questions, 275–76; error of, 277; forms of, 43

Ad misericordiam, 11, 162, 174, 191, 209, 221, 229

Ad populum, 16, 41, 191, 210, 229; fallacy, 41

Ad verecundiam, 115, 162, 191, 197, 209, 269, 273; argument, 16, 17; dialectic analysis, 279

Agenda, 171

Aldrich, Henry, 69

Ambiguity, 61, 85, 256, 292

Amsterdam School, xi–xii, 9, 16

Annals of the Congress of the United States, 39

Appeal: to fashion, 226; to fear, 40, 55, 218–19, 241 to force or threat, 218; to pity, 221–22, 256; to popular pieties, 41; to snobbery, 226; to the gallery, 41, 210; to the glamorous person, 226; to vanity, 226

Appeal to expert opinion, 17, 46, 120, 209, 267; abuses of, 47

Argument: adversary, 28; attack, 27; categories of evaluation, 16; circular, 12, 108–9, 299; counterblaming, 110; deductive, 90; defined, 254–55; dialectic, 6; fallacious, 16, 259, 265; flawed, 264; inductive, 89–90, 96–97, 132; linked, 158; maximally explicit form of, 94; presumptive, 89–91, 93, 276; probative function, 208; reasonable, 16; requirement, 233; strategies, 26, 28–29; support, 27; techniques, 28, 259–62; valid, 82; validity, 75, 235; weak, 16–17, 224, 230, 235–36, 259, 265, 275

Argument du gaspillage, 156

Argument from a threat, 157

Argument from analogy, 60, 135–36; critical questions for, 136

Argument from bias, 153–54, 215; critical questions for, 153

Argument from cause to effect, 140–41; critical questions for, 141

Cost-benefit analysis, 157
Critical discussion, 7, 19, 99, 109, 124, 127, 195, 213, 244, 249, 266, 285; aim, 19–20; basic types, 100; complex, 100; goal of, 13, 20, 99–100, 102, 111, 269; rules, 10–11, 25, 215, 217, 244, 297–301; simple, 100
Critical doubt, 101
Critical persuasion, 20
Critical question, 18, 26, 101

Damer, T. Edward, 60, 296
Debate, 126–27, 170
Deception, 242
De Cornulier, Benoit, 150
De Kruif, Paul, 142
DeMorgan, Augustus, 79–80
Denying the antecedent, 71
Descartes, Rene, 4, 108
De sophisticis elenchis (On Sophistical Refutations), 1, 3, 9, 35, 65, 72, 86, 113, 240, 243, 246
Device of stages, 158
Dialectic, 113, 250
Dialectical argument, 41
Dialectical level of analysis, 201
Dialectical relevance: defined, 163, 168; six factors to prove, 193–94
Dialectical shifts, 197, 254, 285
Dialogue: advice-giving, 187, 284–85, 293; closing stage, 117; context of, 93; criticism of irrelevance, 171, 181, 183, 194; defined, 98, 172; deliberation, 115–16, 284; dialectic, 243; eristic, 111–12, 124, 169, 189, 243, 291; expert consultation, 114; fallacious move, 197; formal, 271; frame based, 239–40, 253; games of, 8, 27–30, 104; goal of, 18, 31, 183–84, 249, 272, 299; information-seeking, 117; initial sequence, 204; maieutic function, 21, 30; normative models of, 98–99; principle of, 179; profile of, 24, 203, 257, 284; quarrelsome, 112; question-answer, 8; question-reply, 180–81, 261; rule violation, 9, 97, 235, 239, 248, 251, 253, 263; sophistical, 111–12; stages of, 99, 276
Dictionary of Philosophy, 221–22, 244–45
Dispute, 100, 171; weakly opposed, 171
Donohue, William A., 105, 219
Duden, 246–47
Dukakis, Michael, 48–49

Ehninger, Douglas, 249
Elements of Logic, 74
Elenchus (reasoning), 2
Engel, S. Morris, 41, 49, 56, 225
Epistemic closure, 150–51
Equivocation, 201, 211, 255
Error: inductive, 52, 89; of insufficient statistics, 89; of meaningless statistics, 89; of reasoning, 245, 291; of unknowable statistics, 89
Essay, 83
Ethotic argument, 152–53; critical questions for, 152
Eubulides, 160
Euclidean geometry, 109
Euthydemus, 111
Evidential priority, 108–9

Fallacia, 240, 250; accidentis, 53; amphiboliae, 79
Fallacy, 239; analyzing an allegation of, 263–64; as a use of argumentation, 14; basic problem with, 2; contextual, 32; criticism of traditional treatment, 14; defined, xi, 7–9, 11, 14, 18, 97, 232–35, 239–42, 245–47, 255–56, 270; elements of, 251; erroneous inference, 2; formal, 75–76, 199, 269; Greek idea, 1; inductive, 199; informal, 78, 266; linguistic shift, 17; modern viewpoint, limitations of, 2; rule violation, 14, 16, 228, 238–40; sophistical tactics, 9–10, 33, 241, 252–53; standard treatment, 35–36; three general problems, 199–200; vs. error in argument, 26
Fallacy: of accent, 63, 86, 201, 273, 296; of accident, 53, 211; of affirming the consequent, 91–94, 96; of ambiguity, 66, 162, 293; of amphiboly, 62, 80–81, 85–86, 201, 211, 273, 294; of composition and division, 64–65, 269; of consequent, 73, 95; of denying the antecedent, 72; of equivocation, 61, 86–88, 273, 293; of four terms, 74; of hasty generalization, 52–54, 77, 89, 123, 211, 269; of interrogation, 177; of irrelevance, 201; of irrelevant evidence, 191; of many questions, 203, 255; of neglect of qualifications, 53, 210–11; of relevance, 66, 162, 191; of rearranging operators, 79; of the consequent, 71; of wrenching from context, 63, 296–97
Fehlschluss, 246–47, 249, 272

About the Series

STUDIES IN RHETORIC AND COMMUNICATION
General Editors:
E. Culpepper Clark, Raymie E. McKerrow, and David Zarefsky

The University of Alabama Press has established this series to publish major new works in the general area of rhetoric and communication, including books treating the symbolic manifestations of political discourse, argument as social knowledge, the impact of machine technology on patterns of communication behavior, and other topics related to the nature or impact of symbolic communication. We actively solicit studies involving historical, critical, or theoretical analyses of human discourse.

About the Author

Douglas Walton is Professor of Philosophy, University of Winnipeg. He received his B.A. from the University of Waterloo and his Ph.D. from the University of Toronto.